A Manual of
Lambing Techniques

A Manual of
Lambing Techniques

Agnes C. Winter
and Cicely W. Hill

The Crowood Press

First published in 1998 by The Farming Press

This edition published in 2003 by
The Crowood Press Ltd
Ramsbury, Marlborough
Wiltshire SN8 2HR

www.crowood.com

British Library Cataloguing-in-Publication Data
A catalogue record for this book is available from the British Library.

ISBN 1 86126 574 3

Illustrations by Jane Upton

Book design by Liz Whatling

Printed and bound in Great Britain by Bookcraft, Midsomer Norton

Contents

To the memory of Tom Winter, a dearly loved husband

Acknowledgements

We would like to thank Judith Charnley and Professor Michael Clarkson
for helpful comments on the script. We are also very grateful
to Jane Upton for so skilfully interpreting our amateur drawings into
clear diagrams which are essential to making the most of the text.

Authors' Preface

How did this book come about? It is based on a combined knowledge acquired over many years working with sheep, with each of us feeling the need for a detailed practical book covering the period approaching, during and immediately after lambing.

One of us (ACW) grew up on a small family farm in the Yorkshire Dales, and from a very young age had a particular affinity for working with sheep. After qualifying as a vet, this interest developed further whilst dealing with day-to-day veterinary work on farms, and included running a small flock of pedigree Wensleydale sheep and their crosses, which experienced many of the problems described in this book! For the last ten years, teaching veterinary undergraduates about sheep, particularly trying to introduce them to the difficulties they will encounter in dealing with problems around lambing time, has shown the need for a detailed book to which they (and others) can refer.

The other one of us (CWH) has spent 25 years in partnership with her husband Peter, working with commercial sheep, Dorset Horns, and a large flock of Poll Dorsets on a frequent lambing system. Experiencing all the problems which face the beginner in the early days, thinking 'How did I tackle this problem last time?', led to the keeping of copious notes and sketches on particular problems and their solutions, which accumulated over some 15 years.

These formed the core around which the book developed, with veterinary and other practical expertise added to produce the kind of book we feel we would have liked to have had to guide us in our inexperienced days. In particular, we want this book to go some small way towards making life better for both ewes and lambs looked after by those who read and use it. We hope we have succeeded.

Agnes C. Winter
Cicely W. Hill

CHAPTER 1

Introduction

Lambing time can be the most exciting and rewarding time of the year in any sheep flock. It can also be the most worrying and potentially the most disastrous if the shepherd is unprepared, inexperienced or handles the sheep unsympathetically. For many years newcomers to caring for sheep traditionally learnt 'on the job' from experienced shepherds. More recently, a new generation of lambers has had hands-on experience at excellent classes run by the Agricultural Training Board. Sadly, many of these training groups have fallen by the wayside, although some training is still given by agricultural colleges and individual veterinary surgeons, and advantage should be taken of these if at all possible. Newcomers may, however, find it difficult to obtain the help of an experienced shepherd or convenient local course. This book aims to help the inexperienced, whether they be agricultural students, veterinary students or someone with a few acres who fancies keeping a few sheep, and perhaps even some experienced shepherds, when presented with a situation in which ewes or lambs appear to be in difficulty around lambing time.

About this book

For those who have little or no experience of sheep, you may read this book and think that things never go right at lambing time! In fact, the vast majority of sheep, providing they are given the opportunity, will lamb perfectly normally and unaided, and the majority of lambs will be alive and healthy. There are exceptions, so it is important to be able to recognise these – for example, some breeds require much more help lambing than others; in some breeds the lambs are less vigorous than others; diseases which cause abortion result in many more lamb deaths than

usual. If you are deciding what breed to buy, it is as well to try to find out about any particular breed quirks before buying. Better still, start with some healthy crossbred ewes which have already reared a crop of lambs so know what the job is about.

As you read through this book, you will be conscious of some repetition, particularly in the sections dealing with difficult births. This is deliberate. We have tried to adopt a logical approach which can be followed in most circumstances, so the same technique is often applied for slightly different presentations; since we envisage the book actually being used in the lambing shed (a bit like a cookery book in the kitchen!), we have repeated instructions, to avoid too much page-turning at what can be a stressful moment for both shepherd and sheep.

It is impossible to cover every conceivable situation which you may encounter around lambing, but most common difficulties are discussed. Our methods are not the only ones which work and doubtless other people will have different methods which they favour. In particular, not every one will agree with the very precise instructions on positioning the ewe during correction of malpresentations; these have been developed and refined over many years and undoubtedly help when faced with a difficult case.

If you are inexperienced, you may find it difficult to handle and position a ewe as instructed during lambing unless you have assistance, but as experience is gained, it should be possible to position all but the largest, most obstinate or wild ewe with the minimum of help. The most important rule you should follow is: **if in doubt, stop or leave well alone and seek the help of**

someone with more experience, usually your veterinary surgeon.

Achieving a successful outcome

The majority of losses in both ewes and lambs take place in the period around lambing. Giving birth and being born are, therefore, high risk periods in the life of the ewe and lamb respectively. Sympathetic and skilful handling at lambing time will reduce mortality and increase the satisfaction of a job well done, as well as maximising the profitability of the flock.

A successful lambing time depends not only upon what happens at the actual time of lambing, but on events several months previously and, having produced live lambs, on keeping the maximum number alive and thriving. Lamb survival is very dependent upon lamb birthweight, which is influenced by litter size and placental size, which are determined in much earlier stages of pregnancy, and by ewe body condition and colostrum and milk production, which are dependent upon nutrition.

In the run-up to lambing, preparation of sheep, accommodation and equipment is essential, together with close observation and prompt action when problems such as prolapses or abortion threaten.

During lambing time itself, the success of dealing with difficult cases relies mainly on five principles.

1. Good preparation of all necessary equipment, which is to hand before delivery is attempted.

2. The use of adequate lubrication, with the highest standard of cleanliness possible.

3. The correct assessment of the presentation of the lamb and appropriate positioning of the ewe, to reduce the pressure of the lamb against the ewe's pelvic bones.

4. The relief of pressure in the birth canal by raising the ewe's hindquarters during some complicated assisted lambings.

5. The careful manipulation of the lamb's head and/or limbs, to produce a streamlined shape, carried out in a gentle and patient manner.

These principles should be observed even in the case of a seemingly small problem, which after thoughtless haste could be turned into a complicated one.

After lambing has taken place, ensuring that lambs receive sufficient colostrum, that they are mothered up properly and that the ewe is producing sufficient milk for all her lambs are crucial factors in carrying through success.

CHAPTER 2

Basic Lambing Equipment

All lambing equipment should be gathered together and prepared for use well in advance of the expected date of lambing. A plastic 'caddy' such as the type used for carrying household cleaning materials, or a 'baby box', is ideal for holding equipment.

Halter

Most experienced shepherds don't bother with a halter, but if you are inexperienced, it will aid in restraining your ewe if other help is not available.

Lambing ropes and snares

A variety of commercial devices is available, many of which have their limitations. It is easy and cheap to make your own ropes and snares (see page 5). Each set should consist of two leg ropes and a head rope and/or snare. For a large flock, or in the case of synchronised lambing periods, it is essential to have numerous sets of lambing ropes available to ensure proper sterilisation between lambings. After use they should be cleaned under running water and soaked in disinfectant, then rinsed well, dried if possible, and put in clean polythene bags ready for next time. At the end of the lambing period they should be dried well before storage. See page 8 for more details about the practical use of lambing ropes.

Bucket

Choose one with a wide base to reduce the possibility of it being knocked over. A bucket with a lid, eg. a nappy bucket, prevents the water spilling if it has to be carried some distance. The bucket should be disinfected between each assisted lambing.

Lubrication

Soap, soapflakes and commercial lubricants are all commonly used. Soapflakes should not be used unless dissolved and prepared as described below. Detergents must never be used, as they sting and irritate if applied directly to a bruised birth canal and can cause the ewe to have difficulty urinating after lambing.

Soap is one of the cheapest and most effective lubricants. A bar of soap (unscented) in hand-hot water will soften on the outside and provide better lubrication on the shepherd's hands and arms than soapflakes. It can be helpful to dig the fingernails into the softened soap to fill them. A cake of soap is also very effective when rubbed well into the lambing ropes before they are inserted into the uterus.

A home-made soap solution can easily be made and stored in a squeezy plastic bottle, which enables the solution to be directed along the hand into the birth canal. Stir one tablespoonful of soapflakes and half a teaspoon of glycerine into 500 ml boiling water. Stir until dissolved, cool slightly (not too much or it will be difficult to pour), then fill the squeezy bottle ready to use.

Of the proprietary lubricants available, the powder type which turns to jelly on contact with water is particularly useful. Some of the jelly types tend to fall off the hand before they get to where they are needed unless care is taken.

In a difficult lambing, where more lubrication is needed but little room is available for the hand, it may be possible to insert a stomach tube past an obstruction and then 'inject' lubricant through it from a large feeding syringe.

Other alternatives include petroleum jelly, liquid paraffin and, in an emergency, lard. If you come across a ewe requiring help unexpectedly and have no lubricant, milking some colostrum onto your hand is better than nothing.

Pieces of dry towelling

These are invaluable for wiping away membrane and fluids from the mouth and nose of the newborn lamb, when every second may count. They should be discarded or washed frequently, to prevent the spread of infection between lambs.

Clean newspaper

This can be laid on grass or bedding to help keep ropes as clean as possible during use.

Prolapse harnesses and/or lengths of soft cotton rope

These should be available so that prolapses can be replaced in the early stages, preferably before the ewe is in a state where she is straining badly.

Navel dressing spray or dip

It is good practice to treat the navel of newborn lambs, particularly in intensive indoor systems, to guard against infection. Spraying the navel cord is not as effective as dipping, unless done very thoroughly, but dips can act as a source of infection and so should be renewed frequently. An antibiotic spray, which can only be obtained from your veterinary surgeon, is one product which is commonly used. Iodine as a spray or dip is an alternative. This should be an alcoholic solution (tincture) rather than an aqueous solution and not more than 2.5% strength. Stronger solutions can cause premature cracking of the attachment of the cord to the skin, allowing in infections. Teat dips for cows are **not** suitable since they are emollient rather than astringent.

Stomach tube and syringe or funnel

This essential piece of equipment allows quick and sure feeding of a lamb which will not, may not or cannot suck. It should be thoroughly cleaned in soapy water and rinsed after use. Prolonged soaking in disinfectant may eventually cause damage to the tube and particularly the syringe.

Supply of colostrum

Surplus ewe colostrum can be stored in the freezer. It is essential for times when a ewe has too little for her lamb(s). This may be because she has not been fed enough during pregnancy, has had mastitis or is ill, or has more lambs than she can feed. Ewe colostrum is best but there are alternatives (see page 68).

Thermometer

This is necessary to check the temperature of lambs, particularly if hypothermia is a possibility; also for checking ewes which appear off colour.

Warming box or heat lamp

For larger flocks, it is well worth investing in a thermostatically controlled warming box. For small numbers of lambs, a fan heater or infra-red lamp can be used, but great care has to be taken not to overheat the lamb or even burn it (see pages 71-2).

Syringes and needles

A selection of disposable syringes including 2, 5, 10 and 50 ml sizes and needles including 16, 18 and 20 gauge x 2.5 cm will cover most eventualities. Dog vaccination syringes and needles are very suitable for treating lambs. Syringes should be safely discarded after breaking off the nozzle. Needles should be changed frequently and safely discarded into a special sharps container (ask your vet again).

Essential drugs

Consult your vet about a supply of drugs such as injectable antibiotic (usually given after a difficult lambing), injections such as calcium borogluconate and treatments for sick lambs, if necessary. A respiratory stimulant is useful and may succeed in helping a lamb which is reluctant to breathe. Most of these are 'prescription only' medicines and your vet will only supply them if familiar with your flock.

CHAPTER 3

How to Make Your Own Lambing Ropes and Snares

Commercially available lambing ropes are mostly made of twisted rope of varying thickness, with a loop at one or both ends; another type is made of rubber which is rather too stretchy. There are also several designs of lamb pullers. Some people use baling twine, but this may be contaminated and can cut into and damage lambs so should only be used with care. We suggest you make your own as instructed below. The designs for the ropes and snare have been refined by trial and error over several years and are superior in use (and cheaper!) to any commercially available.

The type of rope required is synthetic rather than cotton, and braided rather than twisted; it is inexpensive and can be purchased by the metre from shops catering for outdoor activities such as camping, sailing, and climbing. The rope can also be found in some Do-It-Yourself stores. Twisted rope is not so easy to manipulate into position, bits of membrane tend to get caught in the rope during lambing and it is time consuming to clean.

Braided rope has none of the above faults. Most of it is made of an outer sheath round a central core which, when removed, leaves a soft pliable rope ideal for the job. The rope should be about as thick as a standard ballpoint pen (about 7 mm diameter). Choose a bright colour so that the ropes are not easily lost; different colours can be used to distinguish head and leg ropes if desired.

Each set of ropes, ie. a double-ended head rope and two leg ropes, will take in total just under two metres of rope. The only other requirement is some strong thread for stitching.

Double-ended head rope with safety stops

A head rope with a sliding noose (Figure 3.1) is the easiest to place in position on the lamb's head, but it must have a safety stop to prevent it tightening excessively or accidentally strangling the lamb. This type of head rope is ideal, in our experience, especially for very complicated lambings.

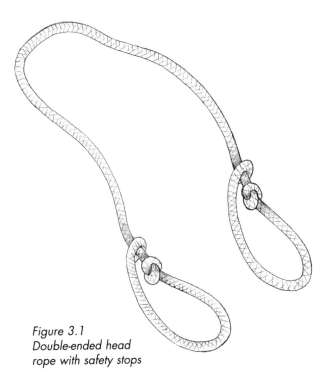

Figure 3.1
Double-ended head
rope with safety stops

1. Take a piece of braided rope measuring 1 m in length.

2. Draw out and discard the central core. The ends of the rope should not be sealed by heat as such action can form hard, sometimes sharp edges. Whipping should not be used. It is too bulky and makes stitching difficult.

3. Tie a loose knot approximately 20 cm from one end (Figure 3.2).

4. Take the end of the rope and double approximately 3 cm over the rope, forming a noose which encloses the loose knot (Figure 3.3).

5. Stitch the doubled-over piece firmly together with strong thread as in the diagram, allowing the rope to **just** slide through the loop, so that when the noose is tightened it will stay in place. Slight fraying at the stitching is of no consequence; in fact it helps to carry lubricant.

6. Repeat the procedure with the other end of the rope.

7. Now position the knots which form the safety stops carefully.

8. Pull each noose out flat, measure the appropriate distance and position each knot. For average to large lambs the flattened noose should measure 11 cm (Figure 3.4). For very small lambs the flattened noose should measure 9 cm (Figure 3.5). Once the knots are in position, they should be pulled really tight. Providing this is done they will not slip during use.

Leg ropes

Two ropes are necessary, each with a sliding noose at one end. Tying one knot at the other end of one rope and two knots at the other end of the second rope will enable each rope to be identified with a particular limb, so avoiding the possibility of the wrong rope being pulled in the case of a complicated malpresentation.

1. For each leg rope take 40 cm rope.

2. Draw out and discard the central core. Do not heat seal the ends of the rope.

3. Create a noose as described for the head rope, but without the safety stop.

4. Stitch the doubled-over piece firmly together with strong thread to form a loop, taking special care that the rope **just** slides through the stitched loop, so that when

Figure 3.2
Making head rope safety stop

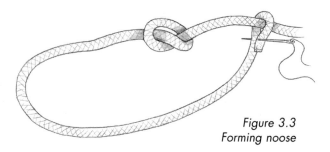

Figure 3.3
Forming noose

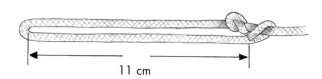

11 cm

Figure 3.4
Position of safety stop medium/large lamb

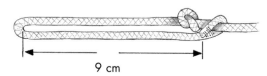

9 cm

Figure 3.5
Position of safety stop for very small lambs

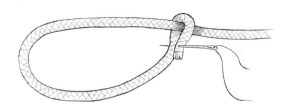

Figure 3.6
Making a leg rope

tightened it will stay in place (Figure 3.6). If this is not done, the rope will loosen and easily slip off the leg.

5. Now tie a knot at the free end of one rope and two knots at the free end of the other rope so that each can be identified.

Plastic-coated wire head snare

Some people prefer a wire snare to a rope for handling the head. One of these (Figure 3.7) can easily be made with a piece of telephone wire or electric flex (the flat type is best, 0.5 to 1mm² size) approximately 1.5 m long. Take care that the wire is not springy; otherwise it will not stay in position on the lamb's head when in use. You will also need the barrel of a 2 or 5 ml syringe, depending on the thickness of the flex, to make a plastic sleeve.

1. Cut off the nozzle end of the syringe (try to get a flexible syringe, not the type that shatters). Make sure the end has no sharp projections.

2. Make sure there are no sharp ends of wire protruding from the cut ends of the cable, then fold the wire in half.

3. Thread both cut ends through the syringe barrel with the flanges away from the folded end of wire.

4. Twist or knot the free ends of the wires together to make a comfortable handle.

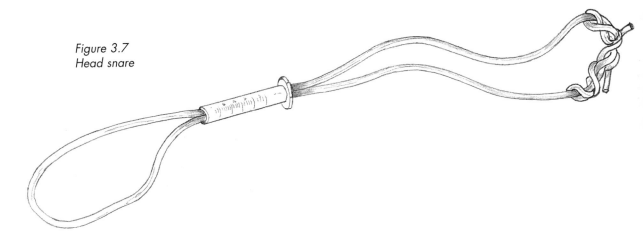

Figure 3.7
Head snare

CHAPTER 4

How to Use Lambing Ropes and Snares

Having made or acquired your lambing ropes and snare, it is now essential that you know how to use them correctly.

Head rope or snare

Although in some cases where the lamb is small, it is possible to manipulate the head with the hand only, the vast majority of difficult lambings require the use of a head rope or snare. The lamb must **never** be grabbed or pulled by the lower jaw, since this can easily be broken, nor should the eye sockets be used.

The home-made rope with a sliding noose and a safety stop is easier to position on the lamb's head and safer in use than other types. A sliding noose should not be used without a safety stop, since it can damage the lamb's head or even strangle the lamb if pulled too hard. A fixed loop head rope or snare is difficult to position on a lamb's head deep in the uterus, for it must be fed under the lamb's chin first and then moved over the back of the head, a difficult manoeuvre in a complicated lambing. The 'lamb puller' type of instrument comprising a stiff shaft with a movable loop at the end cannot be easily used in the uterus because of its rigidity.

Here are two methods of applying the head rope, each favoured by one of us. You may find one method easier than the other. First, assess the size of the lamb. The small noose must not be used on large or medium sized lambs, as it is intended only for very small lambs.

- Lubricate the rope well, lengthen the noose to approximately twice the length of the lamb's head and hold the centre of the noose between

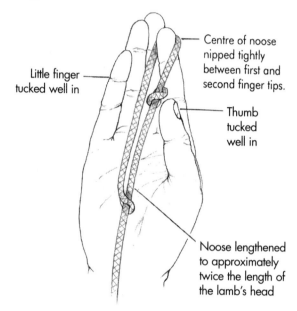

Figure 4.1
Method of holding head rope

Labels on figure:
- Little finger tucked well in
- Centre of noose nipped tightly between first and second finger tips.
- Thumb tucked well in
- Noose lengthened to approximately twice the length of the lamb's head

the tips of the first and second fingers. Keep the thumb and little finger tucked well in to achieve a streamlined shape (Figure 4.1). Move the well-lubricated hand and rope along the nose of the lamb, until the top of the head is reached. Keeping the first and second fingers together with the rope on top of the head, use the thumb to work half the noose round one side of the head and under the jaw bone. Then use the little finger to work the other half of the noose into a similar position on the other side of the head. Pull both ears through the noose and push the noose well down the back of the head (Figure 4.2). Keep the hand and fingers in position on the lamb's head and with the other hand, pull the free end of the head rope to tighten the noose up to the safety stop. It is very important that the safety stop is centred under the chin of the lamb. The hand can now be withdrawn from

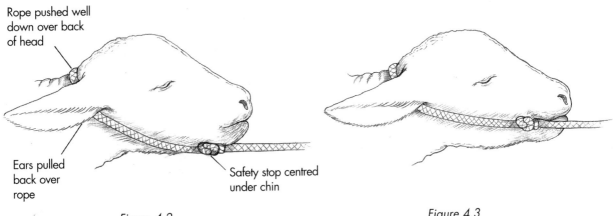

Rope pushed well down over back of head

Ears pulled back over rope

Safety stop centred under chin

Figure 4.2
head rope in position

Figure 4.3
Alternative method of positioning head rope

the uterus, to allow room for the head to be drawn forward and the lamb delivered.

- Alternatively, place the rope over the lamb's head as described above, but instead of placing the noose beneath the jaw, bring one side of the noose between the jaws and, holding it in position with the fingers, pull the free end of the rope with the other hand to tighten the rope so that the knot lies just to the side of the mouth (Figure 4.3). This method has the disadvantage that in opening the mouth, it slightly alters the streamlined position of the head, so may be less suitable for lambs with a large head.

Placing the head rope in position requires a certain amount of dexterity and practice. Some people may find a plastic-coated wire snare easier to manipulate. It is essential that the wire used is not

springy, since that type will not stay in position. Although the wire should be flexible, it should retain whatever position and shape it is placed in.

- With a well-lubricated hand, form the wire into a narrow loop and use the fingers as described above to guide the loop over the back of the lamb's head. The wires can be held near the handle with the other hand which can assist in advancing the wire. When the fingers reach the back of the lamb's head, open the hand slightly to widen the loop and push it well over the back of the head, making sure the ears are flicked out in front of the wire. Hold the loop in position and tighten the wire by pulling with the other hand. Push the plastic sleeve to hold the wire snugly onto the lamb's head (Figure 4.4).

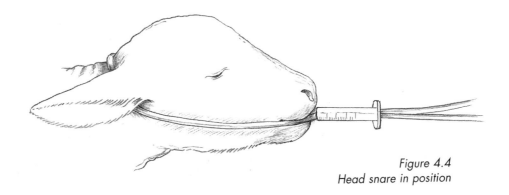

Figure 4.4
Head snare in position

9

Leg ropes

Two leg ropes, each with a plain sliding noose, will be required for the majority of malpresentations. One rope should have one knot and the other rope two knots at the free end, so that each rope can be identified with a particular limb; alternatively make them of different coloured rope.

The leg rope should always be positioned above the fetlock joint (Figure 4.5), **not** just above the hoof, which would risk damaging the hoof; also the rope can easily slide off if placed too low down the foot. It is important that the rope is led off from the **underside** of the foot; otherwise the foot will bend double and may become jammed (Figure 4.6).

Figure 4.5
Leg rope in correct position

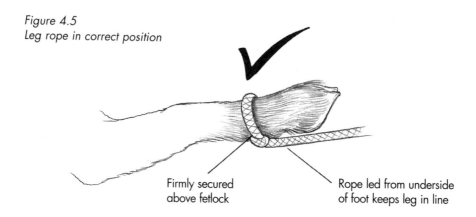

Firmly secured
above fetlock

Rope led from underside
of foot keeps leg in line

Figure 4.6
Leg rope incorrectly positioned

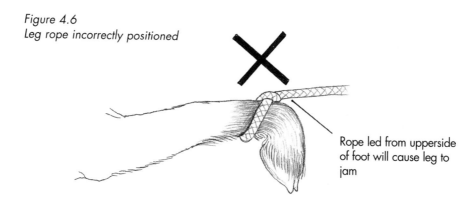

Rope led from upperside
of foot will cause leg to
jam

CHAPTER 5

A Simple Hoist and its Applications

Positioning the ewe

Experienced lambers will have their own views on the best way to position a ewe in order to aid correction and delivery of lambs in difficult cases. There is no doubt that placing a ewe in a specific position – standing, on a particular side, on her back – can make all the difference between struggling to correct a malpresentation and achieving success with relative ease. We make very specific suggestions about positioning the ewe depending on the type of malpresentation you find. With experience it will often be possible to handle and position animals without extra help, although you will need another pair of hands if you are inexperienced or if the ewe is very big or difficult to handle.

With some malpresentations, raising the hind-quarters of the ewe makes correction much easier. By doing this, pressure on the lamb is relieved and it is much easier to avoid damaging the ewe internally whilst manipulating the lamb.

If the ewe's hindquarters are raised, this should be done for absolutely the minimum time necessary to carry out the manipulation and the ewe must be handled gently with care at all stages. She should immediately be lowered if she shows any signs of distress.

Raising the ewe may be done manually by an assistant, but it is back-breaking work and difficult or impossible with large sheep.

The hoist

Raising the hindquarters can be made much easier by the use of the hoist (Figure 5.1). This is basically a short piece of rope with a noose at either end, which is fixed to each back leg above the hocks

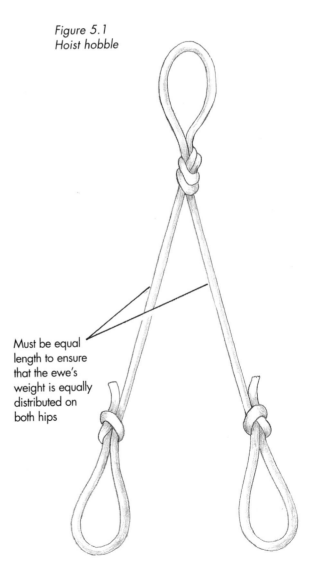

Figure 5.1
Hoist hobble

Must be equal length to ensure that the ewe's weight is equally distributed on both hips whilst the ewe is on her side. She is then rolled onto her back and the hoist is pulled up by another rope slung over a beam (Figure 5.2). This allows the hindquarters of the ewe to be raised (Figure 5.3) by the necessary amount with the minimum of effort and, providing the ewe is not too large or wild, single-handedly, if necessary.

How to make a hoist

1. Take 2 m of rope, the softer the better, and thick enough not to cut into the flesh of the legs (braided climbing rope 1.5 to 2 cm diameter is suitable).

2. Fold the rope in half and tie a simple fixed loop of approximately 10 cm length in the middle (this creates a point where the hoist can be secured temporarily if the shepherd is without assistance).

3. Tie a slip knot at each free end of the rope, making sure each is equidistant from the fixed loop. If these knots are lopsided, the weight of the ewe will not be equally distributed on both hips.

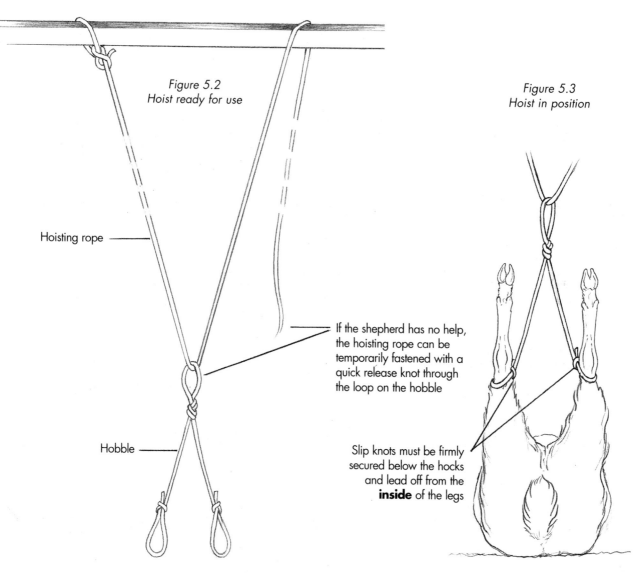

Figure 5.2
Hoist ready for use

Figure 5.3
Hoist in position

Hoisting rope

Hobble

If the shepherd has no help, the hoisting rope can be temporarily fastened with a quick release knot through the loop on the hobble

Slip knots must be firmly secured below the hocks and lead off from the **inside** of the legs

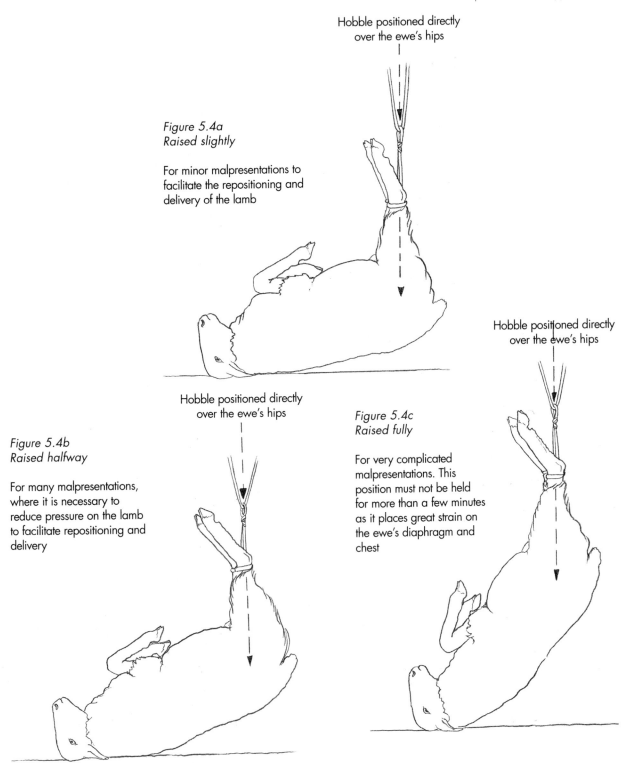

Hobble positioned directly
over the ewe's hips

Figure 5.4a
Raised slightly

For minor malpresentations to
facilitate the repositioning and
delivery of the lamb

Hobble positioned directly
over the ewe's hips

Figure 5.4b
Raised halfway

For many malpresentations,
where it is necessary to
reduce pressure on the lamb
to facilitate repositioning and
delivery

Hobble positioned directly
over the ewe's hips

Figure 5.4c
Raised fully

For very complicated
malpresentations. This
position must not be held
for more than a few minutes
as it places great strain on
the ewe's diaphragm and
chest

Figures 5.4a–c
Situations where the raising of the ewe's hindquarters is required

CHAPTER 6

Essential Notes on the Lambing Ewe

The normal length of gestation is about 147 days, although it can range from 142 to 153 days, with variation depending on breed and litter size. Lambs produced before this time are the result of abortion or premature birth (see p20). Most ewes lamb unaided, especially when allowed to lamb outside in a stress-free environment. However, today many ewes are lambed indoors, which has the disadvantage that it may not allow them to exhibit their full normal behaviour when lambing is imminent. Housing makes things easier from the shepherd's point of view, since the ewes can be observed and caught easily and sheep and shepherd are protected from bad weather, but things can go wrong if management is not of a high standard, eg. mismothering of lambs is common and outbreaks of disease can occur if hygiene is poor.

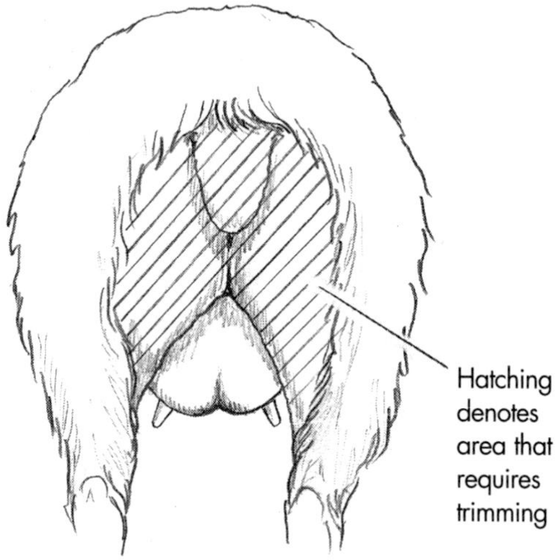

Crutching

Wherever the sheep are to lamb, it is helpful to crutch out the ewes by removing excess wool from around the udder, tail and breech area a few weeks before lambing is due to start (Figure 6.1). In heavily woolled breeds the teats may be surrounded by wool, so the belly near the teats and on the insides of the legs should also be trimmed. The reasons for doing this are:

- improves hygiene at lambing

- makes the first signs of any problem at lambing easier to spot

- enables the lamb to locate the teats more easily

- reduces the risk of lambs swallowing wool which can form into balls in the stomach.

Figure 6.1
Crutching

Hatching denotes area that requires trimming

Hatching denotes area that requires trimming

However, take care not to remove so much wool that the udder is overexposed as this can cause chilling in bad weather, which may lead to acute mastitis in early lactation.

Signs of lambing

A ewe which is about to lamb will generally attempt to find a space on her own in a lambing shed, eg. in a corner. When in a field, she will generally move well away from the rest of the flock, although in the case of a young ewe which has never lambed before, this does not always happen. The ewe will be uneasy, frequently scratching the ground with a front foot, lying down and getting up again. After a while the ewe will start to strain; she will be lying in an unmistakable position with her nose pointing towards the sky (Figure 6.2). The waterbag will emerge (Figure 6.3)

When to examine

The length of time taken for a normal lambing varies according to factors such as age, breed, size of lamb, litter size etc. Interfering too soon can be as harmful as leaving too long, but a gentle examination, with the fingers only, can reassure that all is well if you are concerned about an animal's progress. In general a ewe should always be examined if she has made no progress by one hour after the waterbag has burst. If a normal presentation is felt in the birth canal, and particularly if an unbroken waterbag (like an inflated balloon) is felt, the ewe can be kept under

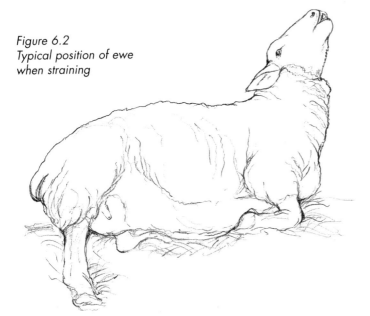

Figure 6.2
Typical position of ewe
when straining

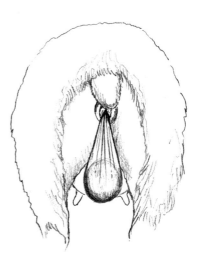

Figure 6.3
Typical appearance of waterbag
before it bursts

and subsequently burst, leaving the thin membranes hanging from the vulva. The ewe carries on straining, repeatedly changing her position to enable the lamb to be pushed down the birth canal. If everything is normal, the lamb's nose together with the two front feet, one either side of the nose, will appear. Once the head and front feet are clear of the vulva the lamb is usually expelled quite quickly. A small percentage of ewes give birth in a standing position, but even then, the ewe's head will unmistakably point upwards as she strains.

close observation and left for a further half to one hour before re-examining more fully. If the ewe is left for too long without assistance, the birth fluids are gradually expelled and the lamb becomes very dry, making delivery increasingly difficult. With some malpresentations, particularly breech, the water bag may rupture within the birth canal, the only sign that this has happened being that the ewe is wet in the region below the vulva and over the back of the udder. Additionally, few attempts may be made to strain, so that these cases are only recognised in the early stages by a particularly observant shepherd.

Moving the ewe when delivery of the lamb must be assisted

In some cases, it is unavoidably necessary to move the ewe a short distance to a more suitable area. If the ewe is lying, make sure her back legs are tucked under her abdomen before trying to get her to her feet. Pulling her up whilst her back legs are stuck out behind her only makes it more difficult and stressful for her and for the shepherd. Pulling and pushing the standing ewe can cause her to panic or start straining, further increasing the problems of delivery. It is much easier and less stressful to hold her head under the chin and then gently push her sideways by putting pressure with the knees alternately on her hips and then on her shoulders.

Moving a ewe which is unable to stand

If a ewe is in an awkward location by a hedge or on sloping ground, moving her sideways is achieved least stressfully by an equal pull alternately on both front legs, then on both back legs.

Placing a standing ewe on her side or back

The ewe should be turned gently. Stand at her side and then, holding the head with one hand, turn it firmly towards the shoulder on the opposite side. At the same time, with the other hand, press down firmly on the loin. She should quietly subside on to the floor and can then be placed fully on her side or back.

Turning a ewe over from one side to the other

Make sure there is ample room, for it is surprising how much space is needed. It is easy to panic a ewe with this manoeuvre; to avoid this, firmly grasp the front and back legs nearest the ground and turn the ewe in one smooth, uninterrupted movement.

Preparation for an examination

If the ewe's vulval area is dirty it should be washed with warm water and soap before beginning the internal examination.

Any examination must be done with a clean, well-lubricated hand (with short nails filed smooth and rings removed), taking great care not to damage the wall of the vagina or uterus. If infection is introduced, or damage caused, the ewe will become ill and may even die. Small hands often have an advantage, although skill is as important as hand size. Hands should be washed with soap or a mild disinfectant and water, and plenty of lubricant applied. Repeated washing and exposure to foetal fluids, particularly in cold windy weather, can lead to the hands and arms becoming chapped. Some people prefer to wear disposable polythene gloves; these help to prevent the arms becoming sore, prevent transfer of infection between sheep, and from sheep to people. They are also useful where access to water is difficult, eg. in remote outdoor conditions. The problem with gloves is that they interfere with sensitivity of touch and may tear if difficult manipulations have to be carried out.

How to examine

The initial examination is best carried out in the position in which the ewe happens to be at that moment, ie. standing or lying. It is essential that adequate lubrication is used throughout any manipulations. It may be possible to do an initial check of what is happening by feeling with the fingers rather than the whole hand. Normal presentations, or some common malpresentations such as nose only or breech birth, may be detected in this way and an appropriate decision to leave or intervene further can be made. If a more thorough examination is necessary, the hand is introduced into the vagina with the thumb tucked well into the palm of the hand and the fingers held tightly together to achieve a streamlined shape. Either hand may be used to carry out the examination. It may be necessary to swap hands during manipulations of the lamb or application of ropes. This is because one hand may become tired, or, particularly when manoeuvring legs, in order to use the hand on the same side as the leg. When feeling the lamb within the uterus, any pressure with the fingers should be applied to the lamb and not to the wall of the uterus which may rupture if handled roughly.

It is quite common for the rectum to contain hard knobs of faeces, which may cause confusion in some cases. The faeces will be passed as the lamb is delivered. If the ewe produces soft faeces as she strains, these should be cleared away from the vulva. Then the hands should be washed and relubricated before carrying on with the examination.

It may be necessary to place the ewe in a particular position when examining certain malpresentations.

Nature of the foetal fluids

There are two types of foetal fluid: thin, urine-like allantoic fluid (in fact it is foetal urine) and thicker, clear amniotic fluid, rather like egg white, in which the unborn lamb floats.

- If the fluids are coloured a bright orange, this shows that the unborn lamb has passed the meconium (foetal dung), and indicates that it may already be in some distress. Prompt action should be taken to deliver the lamb(s).

- If an abnormal amount of bright red blood is present, great care must be taken in examining the ewe in case of internal damage. It may also indicate that the umbilical cord has already ruptured.

- If dark brownish fluid is seen, particularly with an unpleasant sweetish or obvious foul smell, this probably indicates the presence of a dead lamb. At this stage the lamb will feel dry and its wool may have started to come away, depending on how long the lamb has been dead. In very delayed cases, the lamb may be blown up with gas making delivery very difficult. In the majority of cases where the lamb is dead the ewe will require assistance.

The same principles required to deliver a live lamb are applied if the lamb is dead, with extra care that the vagina and uterus are not damaged and opened to infection. Before attempting delivery, make sure that a bucket or plastic sack is to hand so that the dead lamb can be immediately removed to avoid contamination of the lambing area. Dead lambs should be burned or buried as soon as possible.

CHAPTER 7

The Sick In-lamb Ewe

The heavily pregnant ewe is vulnerable to a number of diseases and disorders. The most important and common are abortion and prolapses, which are dealt with separately in Chapters 8 and 9.

Ewes which are most likely to become ill are those which are too thin or too fat. The feeding of ewes during late pregnancy, and the technique of condition scoring to monitor body condition, are outside the scope of this book, but are vital parts of management for a successful lambing time. If ewes enter the last 6 to 8 weeks of pregnancy in good, but not fat, body condition (minimum score 2.5, maximum 3.5, depending on breed), are fed with good-quality roughage and concentrates appropriate to expected lamb numbers and maintenance of body condition, are handled quietly and are allowed to exercise, they should usually have few problems. Sudden changes in housing or food, bad weather and rough handling during vaccination or drenching can precipitate diseases, particularly twin lamb disease and hypocalcaemia.

Here are some of the most common possibilities if one or more ewes become sick at this stage. There are many other reasons for sheep to be ill, so you should always call your vet if several sheep are affected, if you are unsure of the cause or if sheep fail to respond quickly to any treatment you give. Delay is likely to result in a dead ewe and dead lambs.

The ewe is off her food but her temperature is normal (39 to 40°C)

The most likely cause is the early stages of twin lamb disease. If the ewe hasn't already got this and refuses to eat for a day or two, she soon will have it! It's particularly important that any ewe which does not eat is looked at immediately. Don't wait to see how she is tomorrow as that may be too late. Drench with some twin lamb disease remedy (get this from your vet), offer a little palatable, high-energy food such as sugarbeet pulp (soaked with water) and flaked maize (not more than 0.5 kg at a time to avoid stomach upsets) and give some top-quality hay. If the ewe is housed, turning out to grass for a few hours may help to regain the appetite. If there is no improvement by the next day, get the vet to see her.

Other possibilities include:

- acidosis (an upset rumen) caused by sudden changes in feeding (is she scouring?)

- early stages of hypocalcaemia (see page 19).

The ewe appears blind

If the ewe is dull, not interested in food and appears to be blind, she is probably suffering from more advanced pregnancy toxaemia. The outlook in these cases is poor. Your vet may give glucose into a vein, and may consider giving an injection to try to induce lambing.

The ewe has a high temperature (over 41°C)

This is the result of an infection; possibilities include:

- pasteurella pneumonia – is she breathing fast, coughing and has she a dirty nose?

- some types of abortion (especially salmonellosis) – is there a smelly vaginal discharge?

- acute mastitis – is the udder swollen and painful?

Treatment will depend on the cause. If there is nothing obvious to see, an injection of antibiotic, given after consultation with your vet, may be sufficient. If several ewes are affected or there are signs of impending abortion, or other obvious problem which you cannot deal with, you should consult your vet in any case.

The ewe is unsteady on her feet

The most likely reason for this appearance in late pregnancy is hypocalcaemia. The condition of the ewe will gradually worsen over several hours if not treated. It often occurs after handling, housing or bad weather and several ewes may be affected. Try giving 50–100 ml calcium borogluconate 20% by injection under the skin over the ribs. Warm the bottle of calcium by putting it in a bucket of warm water. Use a large syringe with a 16 gauge needle so that you can inject quickly. Rub in well as this injection is painful. A ewe with straightforward hypocalcaemia should respond quickly and be more or less back to normal in a couple of hours.

The ewe cannot stand

This can be the result of, among other things, advanced twin lamb disease, hypocalcaemia, acidosis, septicaemia from pneumonia, mastitis or dead lamb inside her. Try an injection of calcium, as above, if there is nothing obvious to see, but do be prepared to call the vet if there is no change within a couple of hours.

The ewe is flat out

The end stage of all the above diseases will give this appearance. The outlook is very poor unless, miraculously, it is straightforward hypocalcaemia and the ewe responds to calcium.

If the ewe is very twitchy or having fits, it could be hypomagnesaemia (this is far more common a few weeks into lactation). Unfortunately, these cases die very quickly, often before it is possible to give treatment. If you have some magnesium sulphate 25% injection, you can give 50 ml under the skin in the same way as giving calcium.

One other possibility is listeriosis, which is an infection of the brain. Are the ewes being fed silage? This disease is often, but not exclusively, associated with poor-quality silage. Are any others affected? Look for early signs such as depression, a drooping ear on one side, drooling saliva, circling. Response to treatment is very poor unless given very early in the course of the disease. Remove any suspect silage and get advice from your vet.

CHAPTER 8

Abortion

Be aware that most of the infectious causes of abortion can cause human illness, so take great care in handling an aborting sheep or aborted lambs. Wear protective clothing which can easily be washed, disposable gloves, use plenty of disinfectant and dispose of any aborted foetuses or afterbirths (other than any needed for laboratory tests) by burying or burning. Pregnant women should not have any contact with lambing ewes, even indirect (eg. with overalls needing washing), particularly if abortions are occurring. Weak lambs should not be brought indoors: have a special hospital area for them in a building away from the house.

Abortion is the birth of lambs before the end of the normal gestation period; aborted lambs are usually born dead or die soon after birth. Lambs born just before their normal time are usually termed 'premature' and although they may be alive at birth it can be difficult to keep them alive because of the immaturity of their body systems, particularly the lungs. They also have little stored energy reserves, so are very vulnerable to hypothermia.

Most sheep flocks experience the occasional abortion or premature birth for no apparent reason: in about half of the samples sent in to laboratories from aborting ewes no infectious cause is found. **But** the first abortion **may** be the first of several due to some infectious cause so it is wise to treat any seriously.

What to do if a ewe aborts

Most abortions are found when the lamb(s) have already been produced, but sometimes a ewe is found in the process of aborting. If the expected date of lambing is known, there is no difficulty realising what is happening. If the date is not known, poor udder development together with difficulty in performing an internal examination because the birth canal has not enlarged and relaxed should point to the possibility of an abortion occurring.

Here is some advice to follow if a ewe appears to be going to abort or has already done so.

- Remove the ewe to an isolation pen away from the other sheep.

- If the ewe has not yet produced any lamb, carry out a gentle internal examination, wearing disposable gloves.

- If the ewe's birth canal is tight, do not force your hand in as this could seriously damage the ewe – the lamb is likely to be dead so there is no rush to remove it. Most ewes will pass the lamb eventually (it may take a day or so), or at least push it far enough for you to grasp, to complete removal. If in doubt, let the vet have a look.

- Use plenty of lubrication if you are able to get hold of the lamb to remove it.

- Put the dead lamb plus afterbirth (important!) in a clean polythene bag.

- Remember there may be one or more further dead lambs, but do not try to remove them immediately unless you are sure you can do it without damaging the ewe. Allow her time for uterine contractions to push the lambs within reach.

- Clean up the area where the ewe was trying to lamb. If indoors, remove contaminated bedding if practical, then disinfect before covering the area generously with fresh straw.

- Keep the ewe in isolation until a diagnosis is made and mark her so you don't forget which she is.

- Contact your vet, who will advise on sending or taking specimens for laboratory examination to try to determine the cause. The most useful specimen is the afterbirth, particularly a piece with a cotyledon (button) attached, which should be sent together with the aborted lamb(s). The lab should be able to give you some information within 24 hours (perhaps even sooner if the news is bad), although some tests take longer.

- Even if the ewe has milk, do not foster any lamb on to her, at least until the results of the tests are available. This is in case the cause of abortion is infectious to the lambs.

If you are unsure what to do at any stage or the ewe is obviously ill, contact your vet for advice.

Common infectious causes of abortion

Here is some advice if one of the common infectious causes of abortion is diagnosed. Whatever the cause, continue with good hygiene measures: collect and destroy afterbirths, clean out and disinfect lambing pens between ewes, wear disposable gloves and protective clothing which can be easily washed.

Enzootic (chlamydial) abortion

- This infection is particularly dangerous to pregnant women and anyone with a weakened immune system (eg. on steroid or immunosuppressant treatment).

- This is infectious to other sheep, so keep any affected ewes away from others until any discharges cease (about 3 weeks).

- Foster on only male lambs which will go for slaughter.

- Consult your vet about giving ewes yet to lamb an injection of long-acting antibiotic (oxy-tetracycline). This will reduce the number of ewes aborting and should help to save some lambs.

- Consult your vet about using vaccine before the next tupping season.

Toxoplasmosis

- Aborting sheep are not infectious to other sheep so there is no need to separate; it is safe to foster on lambs.

- Ewes aborting as a result of toxoplasmosis are not very infective to people, but the usual precautions should still be taken. Toxoplasma does infect humans, but the infection generally comes from cats (emptying litter trays, contaminated soil) or from eating undercooked meat.

- Cats, particularly young ones, are responsible for spreading the infection (in their faeces) so keep all feedstuffs protected from them. This doesn't mean eliminating all cats, but it is a good precaution to have a stable population of adults without lots of kittens around.

- There is no really effective treatment to give ewes yet to lamb, unless they are a long way off perhaps, so you will have to grin and bear it this time.

- Keep ewes which have aborted – they are now immune.

- Consult your vet about using vaccine before the next tupping season.

Campylobacter

- This is likely to occur as a 'storm'. Ewes which meet the infection become immune, so mix aborted ewes with others, but only after lambing.

- Consult your vet about antibiotic treatment for those yet to lamb.

- It is better not to foster on lambs as the infection may cause them to scour.

- Don't worry about next year – it's unlikely to happen again.

- Protect troughs and food from bird droppings.

- There is a risk of human infection – diarrhoea is the result – so take care.

Salmonellosis

- Some types of salmonella cause severe illness in both sheep and people, so great care is needed.

- Take veterinary and medical advice.

CHAPTER 9

Prelambing Prolapses and Ruptures

Prolapse of the vagina and cervix

Prolapse of the vagina and/or cervix is common in some flocks, yet may never happen in others. There is no one known cause, although overfatness, lack of exercise, carrying triplets, type of diet, docking the tail too short and various other suggestions have been made.

A prolapse usually first appears as a 'blob' of pink tissue (Figure 9.1), which hangs from the vulva when the ewe is lying , but disappears back into place when she stands up. This may go on for several days, causing no concern to the ewe. If simple action is taken at this early stage, by applying a commercial or home-made harness (truss), the prolapse should not become a problem and the ewe should carry on as normal to lambing, although she should be watched carefully as the harness will need to be taken off when lambing is imminent.

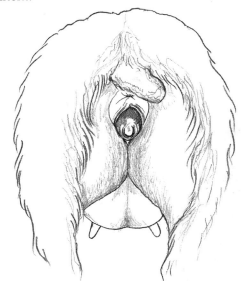

Figure 9.1
Typical appearance of prolapsed cervix in early stages

Applying a home-made truss (Figure 9.2)

1. Take a piece of soft cotton rope about 1 cm diameter and about 3 m long. Make a girth by passing it around the ewe's chest behind the front legs and knotting firmly in the midline behind the shoulders. Make sure that the two free ends are of equal length.

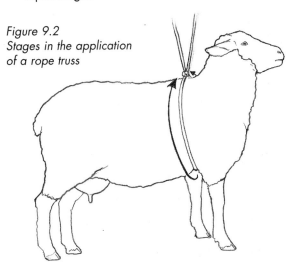

Figure 9.2
Stages in the application
of a rope truss

2. Bring the two free ends together along the backbone, knotting together level with the hip bones.

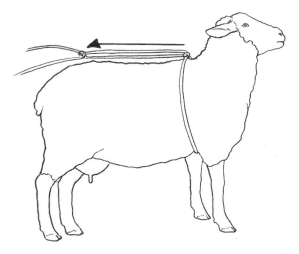

3. Separate the free ends, taking one each side of the tail to lie one each side of the vulva.

4. Cross the ropes just beneath the vulva, taking each free end between the udder and the opposite back leg.

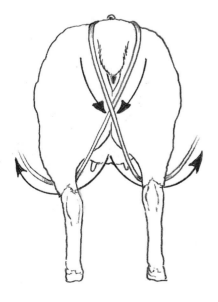

5. Bring each end out in front of the leg.

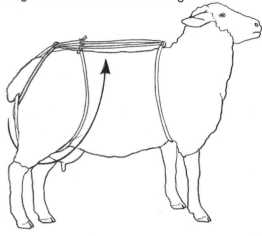

6. Firmly tie each end to the pieces running along the backbone. Adjust the final knots so that the whole harness fits without any slackness, particularly the sections supporting the vulva.

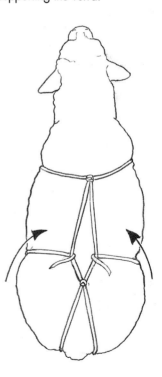

If this type of rope harness, or a well-fitting commercial harness, is applied in the early stages, the prolapse can usually be prevented from getting any worse.

Prolapse accompanied by straining

A simple small prolapse can rapidly turn into a complicated and potentially life-threatening condition when the ewe starts straining. This happens when the urethra, the tube through which urine passes from the bladder, becomes kinked, the bladder fills, the ewe strains to try to produce urine, and a vicious circle of straining is set up. The ewe may also become constipated and strain to pass faeces which adds to the problem (a single dose of about 100 ml liquid paraffin often helps relieve this). The prolapse becomes much bigger, reddened and swollen (Figure 9. 3). The surface becomes dry and inflamed, and may become damaged and infected. As the ewe strains even more, parts of a lamb may be pushed into the prolapse, making things even worse. **Any prolapse should therefore never be neglected.**

Figure 9.3
Appearance of a severe prolapse

What to do if the ewe is straining

1. Examine the prolapse carefully. If it is damaged or if the ewe is straining severely, consult your vet. It is possible for the vet to put an injection into the spinal canal which will make replacing the prolapse easier, make the ewe more comfortable and control straining for 24 hours or so. It also removes pain to allow stitching if this is necessary. Your vet may even decide to induce the ewe to lamb if the straining is going to be difficult to control.

2. Look at the cervix, the entrance to the womb. Check for any sign of foetal membranes or water-bag (Figure 9.4). If present, the ewe must lamb within a few hours; otherwise infection will get into the womb and the end result is likely to be dead lamb(s) and dead ewe. Often in these cases the cervix is only partially dilated. Do not try to lamb the ewe yourself unless you are experienced – forcing the cervix open is likely to cause severe bleeding and shock.

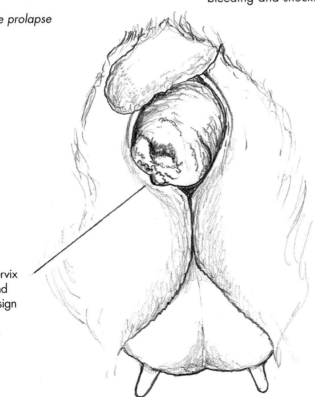

Note the cervix is closed and there is no sign of foetal membranes

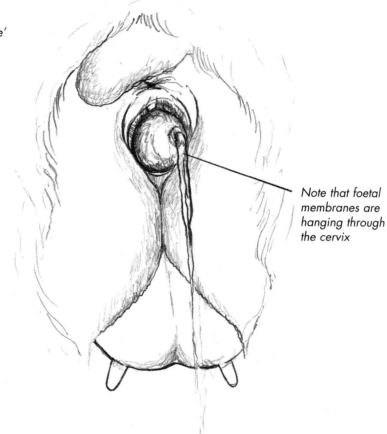

Figure 9.4
'Open prolapse'

Note that foetal membranes are hanging through the cervix

3. If the ewe is not straining too badly, the prolapse is not damaged and the cervix is closed, you can attempt to replace the prolapse.

4. Gently clean the surface of the prolapse with warm water. Do not use strong disinfectants which can cause irritation.

5. Warm the prolapse by applying warm water with a clean piece of towelling – this will help to reduce straining when it is replaced.

6. Raise the hind end of the ewe slightly – use the hoist or put over a bale of straw.

7. Apply plenty of lubricant.

8. Cup your hands gently over the prolapse and, using the palms of both hands, not the fingers, carefully push the prolapse back in during the interval between the ewe's strains. (There will be a gush of urine as the urethra straightens).

9. Keep the hand in place until the urine has been passed and straining is reduced.

10. Gently lower the ewe and quickly put on a well-fitting harness or truss to prevent recurrence. If the ewe carries on straining, stitching may be necessary.

Care until lambing

- Make sure the ewe gets some exercise every day. A lazy ewe which lies around is more likely to try to prolapse again.

- Feed several small meals a day rather than all at once.

- Keep a close watch for signs of lambing; loosen the harness or remove any stitches when membranes or waterbag appear and be prepared to give assistance if necessary.

- Mark the ewe so that she can be identified for culling as there will almost inevitably be a repeat performance next lambing time.

If you particularly wish to keep a ewe which has prolapsed then lambed successfully, she must be identified and watched carefully in the later stages of pregnancy, so that a harness can be applied as soon as the first sign is noticed. Whether the ewe or her lambs should be bred from is another issue, but at present there is not enough evidence of a hereditary tendency to firmly advise against doing so.

Other methods of treating prolapses

There are a number of other methods of treating prolapses, but all have disadvantages compared with the external harness:

- Intravaginal devices ('spoon') are used quite commonly but are not effective if the ewe is straining hard. There is also the risk of introducing infection into the tissues of the vagina and cervix. We do not use them.

- Wool tying is an old-fashioned method in which pieces of string are tied to chunks of the ewe's wool. The strings are then tied together, pulling the wool tight across the vulval region. The method can work if the ewe has a sufficiently long tight fleece, but too often ends up with the chunks of wool being pulled out and the fleece being spoilt. It also leads to soiling of the area beneath the tail as the ewe cannot pass urine or faeces properly.

- Various metal devices, safety pins etc. are very painful for the ewe, likely to cause tearing and should never be used.

Stitching

This may need to be used in problem cases. It should be done by a vet if at all possible, who will give an injection into the spinal canal to reduce straining and provide pain relief. Although a method of stitching is described below, you should only carry this out if you have received some instruction from your vet and are confident that you can perform the procedure without injuring the ewe or causing unnecessary pain.

There are several patterns of stitching in common use. Views differ about whether stitching through the wall of the vagina causes problems or not. The method described does involve stitching through the wall of the vagina, but incorporates supports on each side of the vulva which prevent tearing of the stitches if the ewe does strain.

You will need:

- a large sharp needle such as a Buhner needle (Figure 9.5) which has the eye at the sharp end and a handle at the other end and is designed to do this job. It should be boiled before use

Figure 9.5 Buhner needle

- about 60 cm suture material (this can be 6 mm nylon tape). This should be put in hot water with a few drops of mild disinfectant and then covered with antiseptic cream before use

- some short lengths of polythene or gas tubing through which the suture material can be threaded

- clean scissors

- local anaesthetic; about 1 ml can be injected at the site of each stitch Figure 9.6).

The direction in which the needle is inserted depends on the type. The method described below is for a Buhner needle. If an ordinary needle is used, the pattern of stitching to produce the same end result is more straightforward. The Buhner needle has the advantage that it is inserted from the inside of the vulva outwards at each step, which avoids accidentally catching the wall of the vagina if the ewe strains during the procedure. The stitches should not be placed less than 2 cm from the edge of the lip of the vulva. One continuous stitch is usually adequate, placed so as to leave a gap which will admit one finger at the bottom end of the vulva so that the ewe can pass urine. Stitching should be carried out with the ewe's hindquarters in the raised position, to reduce the chance of the prolapse being pushed out during the stitching procedure (Figures 9.6 and 9.7).

1. Insert the needle through the lip of the vulva just below the top from inside to outside.

2. Thread with the suture material, pulling about 10 cm through the eye.

3. Draw the needle back, pulling the suture with it.

4. Insert the threaded needle through the opposite lip of the vulva, from inside to outside.

5. Unthread the suture material from the eye and hold the end firmly.

6. Withdraw the needle. Adjust the suture material so you have pulled through just under half the length (be careful you don't pull it all through!). Slip a piece of plastic tube (about 4 cm long) over the loose end of the suture material.

7. Insert the needle from inside to outside the lip of the vulva, nearer to the bottom end, so that the needle is 4 to 5 cm away from the first site, making sure there will be sufficient room for the ewe to urinate when the suture is tied. Thread the needle so that the piece of plastic tube remains in place over the suture material.

8. Draw the needle back, pulling the suture with it. Insert the threaded needle through the opposite lip of the vulva, from inside to outside. Unthread and remove the needle.

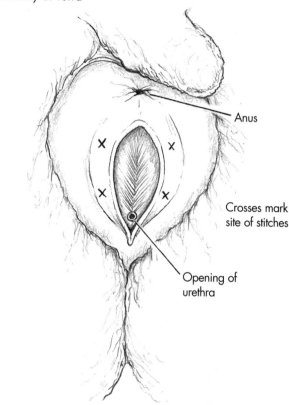

Figure 9.6
Anatomy of vulva

Anus

Crosses mark site of stitches

Opening of urethra

9. With both ends of the suture now on the same side of the vulva, slip another piece of plastic tube over one end and tie the ends together in a knot (or a bow which allows for readjustment), so that the lips of the vulva are just drawn together, not too tightly. The pieces of plastic tube act as a support either side of the vulva and prevent the stitch from being torn out. Cut off the surplus suture material, leaving a 'tail' of at least 5 cm.

Hazards associated with stitching

• The suture is placed too close to the opening of the urethra, preventing passing of urine. This opening is just inside the vulva on the floor of the vagina, so keep any stitches well clear of this area unless you have been specifically instructed by your vet (there is a type of buried purse string suture called a Buhner suture

Figure 9.7
Method of stitching the vulva

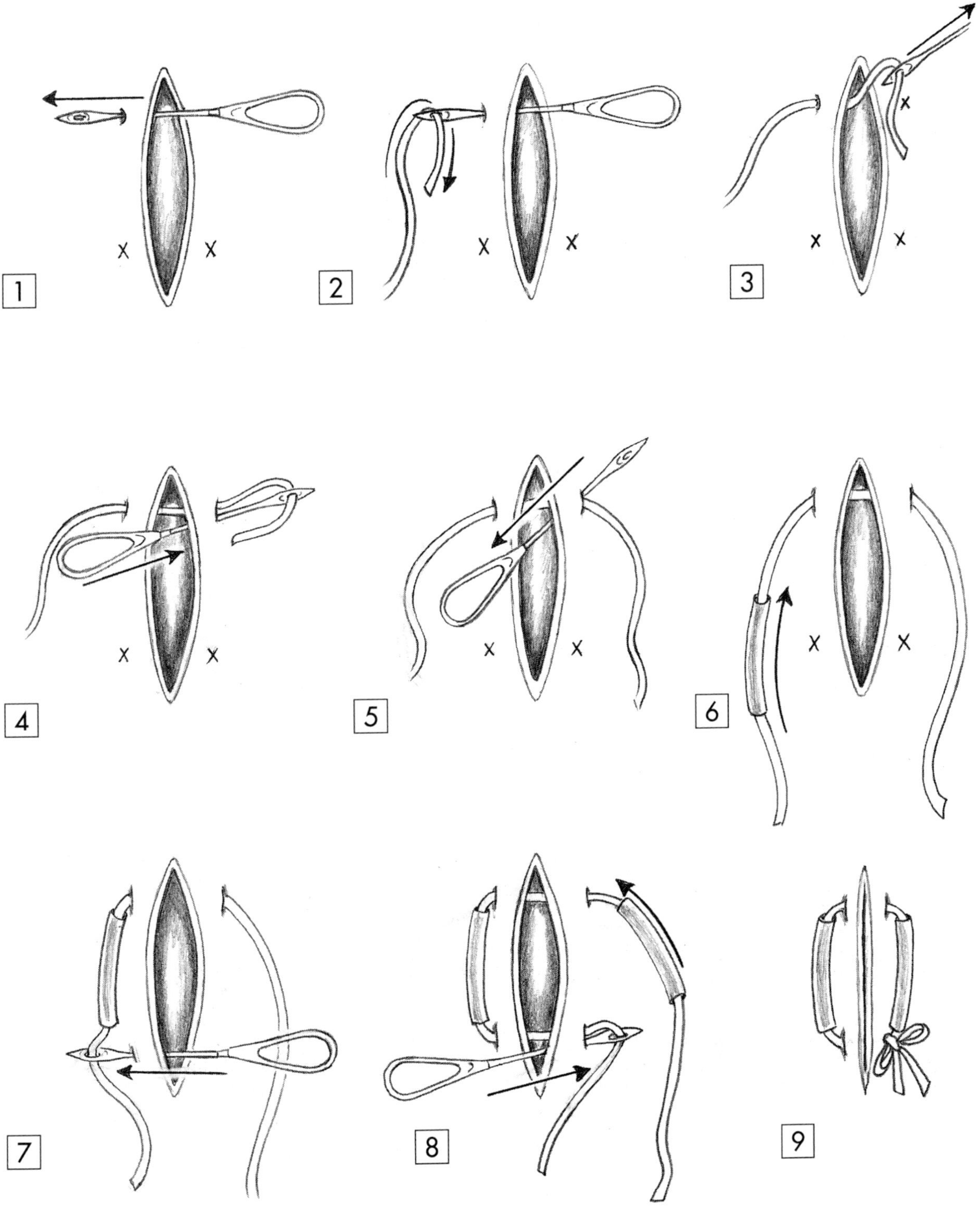

which is inserted starting below the vulva).

- The ewe carries on straining and the sutures are torn through the lips of the vulva. This is most likely if the suture material is too thin, or the sutures are placed too near the lips of the vulva. If the ewe does carry on straining after placement of sutures, you should get her seen by your vet, who will give an injection into the spinal canal to control the straining, or may induce her to lamb.

- The vulva swells becoming oedematous around the stitches, which then may tear out. This often happens if the stitches are tied too tightly, or if they are not supported as described above.

Remember that a neglected prolapse can lead to the death of the ewe and lamb(s). If in doubt consult your vet.

Prolapse of the intestines through the vagina

This is a sporadic calamitous event, which almost inevitably results in the death of the ewe and lambs. Often the ewe is found dead with a mass of intestines spread out behind her. If a ewe is found still alive in this condition she should not be moved, but should be humanely killed as soon as possible. The vet may attempt an emergency caesarean operation to get the lambs out, but as this tragedy usually happens a few days before lambing is due, the lambs are premature and rarely survive. The cause is not known.

Rupture of the abdominal muscles

This occasionally happens in late pregnancy, presumably because of the weight of the lambs, weakness of the muscles or as the result of pushing at overcrowded feed troughs. The end result is that one side of the abdomen (usually not both) 'gives way' and makes the ewe very lopsided (Figure 9.8). Whether the ewe can be left to go through to lambing depends on how bad the rupture is – slaughter may be the only answer in severe cases. If the ewe can manage to get around, she should be kept on her own, within sight and sound of other sheep, in a pen with no projections which might damage the abdomen, and watched carefully for signs of lambing. She will almost certainly need assistance, as the uterus containing the lambs will have fallen into the bottom of the abdomen creating a sort of 'S' bend, which prevents normal birth taking place. Delivery of the lambs can be helped by placing the ewe flat on her back, which allows the uterus to fall into a more normal position giving easier access to the lambs.

Figure 9.8
Rupture of abdominal muscles

CHAPTER 10

Is the Ewe Ready to Lamb?

One of the most worrying aspects for the inexperienced shepherd dealing with lambing ewes is how to tell if a ewe is ready to lamb or not, and whether she requires help. It is quite normal for a string of thick clear mucus to hang from the vulva in the last few weeks of pregnancy – this comes from the neck of the womb – it is the plug of mucus which keeps the entrance to the uterus closed. This should not be confused with foetal fluids or placenta. It is also quite common for heavily pregnant ewes to have difficulty getting comfortable when lying down, so they may have several attempts to do so. They may also strain when defaecating. All these may give rise to false alarms.

With a straightforward lambing in a ewe which has lambed previously, the whole event – uneasiness, separation from the flock, pawing at the ground, straining, appearance of the waterbag, delivery of the lamb – may be over within an hour or so. With a yearling producing what may be a large single as her first offspring, the whole process can take several hours, mainly because much stretching of the birth canal must take place before the lamb can be delivered. Interfering too early, particularly by pulling on the front legs before the head of the lamb engages in the pelvis, can turn a slow but normal lambing into a very difficult malpresentation. On the other hand, dismissing the 'uneasiness but no sign of a waterbag' common with a breech presentation as 'not yet ready to lamb' will lead to a dried-up dead lamb which takes all the experienced shepherd's skills to deliver.

Another situation which worries the beginner is the ewe with a prolapsed vagina or cervix which strains repeatedly. Is the ewe not yet ready to lamb, in which case the prolapse must be replaced, or has lambing started, in which case the lamb will die if it is not born within a few hours.

The key to dealing with these situations is careful observation, quiet handling and gentle examination, sufficient to be able to decide whether:

- the ewe is not ready to lamb and all is well

- the ewe is not ready to lamb, but needs treatment which you can provide

- the ewe is in the process of lambing, but all is well and more time can be allowed

- the ewe is ready to lamb, but there is a problem which you, the shepherd, can cope with

- there is a problem, but you cannot determine the cause or it is too difficult for you to attempt to deal with – seek veterinary attention.

Do not be afraid to carry out a gentle vaginal examination to decide which applies.

- With a vaginal or cervical prolapse where the ewe is not ready to lamb, the folds of cervix around the neck of the womb are closed, a bit like cabbage leaves or a rose. There may be some thick clear mucus present as described above. If a waterbag or foetal membranes are hanging through the cervix, the sheep needs to lamb within a couple of hours. If the cervix does not fully dilate in this time, you are probably dealing with a case of ringwomb following the prolapse (see page 36).

- Internal examination of the vagina in a ewe not ready to lamb will show that the vagina feels dry, with no signs of foetal fluid or membranes. It should be possible to feel the closed entrance to the cervix. It may be possible to feel the legs or head of the lamb through the thin tissues at the side of the cervix – **do not** mistake this for a ringwomb (see page 36) and do not force your way into the uterus.

- If the rounded end of a waterbag is present, be careful not to burst it; leave the ewe for a further 30 minutes or so before checking again.

- If the nose and feet of the lamb can just be felt, again allow more time. If back feet are felt, it is probably better to help the ewe as she will usually have difficulty or take too long unless the lamb is small.

- If the nose is visible and the lamb's tongue is swollen, help should be given straightaway.

- If an obvious malpresentation is felt, make sure you have all the necessary equipment available and then proceed to give help as indicated in the later chapters of this book.

Above all, remember **if you are unsure, get help from your vet.**

CHAPTER 11

Essential Notes About the Lamb to be Delivered

The majority of lambs are born in anterior presentation (Figure 11.1), that is with the nose and front feet appearing first, pointing towards the ground. A smaller number are produced in posterior presentation, that is with the hindlegs appearing first, with the toes the other way up. It is rare for lambs to be born upside down.

Figure 11.1
A normal anterior presentation

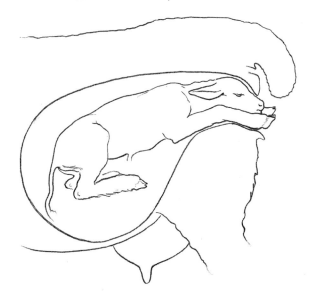

Distinguishing front legs from hind legs

This is a most important part of becoming skilful at lambing. One of the worst mistakes that can be made is to assume that if only two feet are present with no sign of a head, they must be back feet, and to pull them without further examination. If they are front legs and the head is slightly displaced, pulling the legs will turn what can already be a difficult lambing into an almost impossible one, for

it will force the head of the lamb to be turned far back as well as upsetting the normal positioning of the lamb's shoulders and elbows.

It should be possible to get an idea which way the lamb is coming by looking at the feet if they are in view (providing the lamb is not upside down):

- front feet appear with the soles facing down and the dewclaws on the underside of the leg (Figure 11.2)

- back legs appear with the soles facing upwards and the dewclaws on the upper side of the leg (Figure 11.3).

Figure 11.2
Usual appearance of front feet

Figure 11.3
Usual appearance of back feet

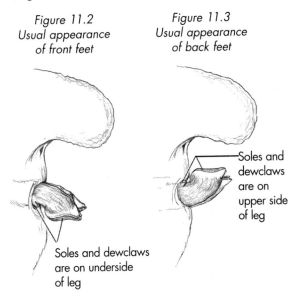

Soles and dewclaws are on upper side of leg

Soles and dewclaws are on underside of leg

Confirmation requires examination, usually by feeling, of the lower parts of the leg (Figure 11.4). Here are three ways of remembering how to do this. The first is useful if the lamb is very tight and it is difficult to bend the joints.

Figure 11.4a
Identification of legs

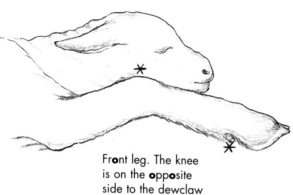

Front leg. The knee
is on the **opposite**
side to the dewclaw

Figure 11.4b
Identification of legs

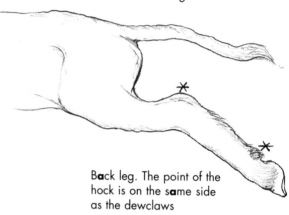

Back leg. The point of the
hock is on the **same** side
as the dewclaws

- with a fr**o**nt leg, the knee is on the **opposite** side to the dewclaws

- with a b**a**ck leg, the point of the hock is on the **same** side as the dewclaws

- in **front** legs the knee and the fetlock joints bend in the same direction (like the letter C at the **front** of the alphabet)

- in **back** legs the hock and fetlock joints bend in opposite directions (like the letter Z at the **back** of the alphabet)

- front legs are like the human arm – the knee and fetlock joints bend in the same direction

- back legs are like the human leg – the hock and fetlock joints bend in the opposite direction

Occasionally the legs may feel abnormal or the joints will not move (are ankylosed). This often suggests that the lamb is deformed and delivery may be complicated.

Manipulating the lamb

When using the hand to pull on the lamb's head, always place it over the top of the head with the fingers behind each ear (Figure 11.5). Never pull with the hand under the chin or by holding the bottom jaw, as this may result in a broken jaw.

When pulling on the legs, do not grip the hoof as this may cause damage to the foot. The leg should always be grasped above the fetlock, with ropes applied also above the feltocks (see Figure 4.5 page 10) if the feet are very slippery and difficult to get hold of.

Direction of pull

The lamb should always be delivered in a curve towards the ewe's hocks, with the exception of a lamb which is presented on its back (Figure 11.6). This is rare. Where the lamb is tight, for example with one leg back or a large lamb coming backwards, it is helpful to twist the lamb slightly at the same time as pulling. This helps to prevent the shoulder or ribs from becoming locked against the brim of the pelvis.

Figure 11.5
Position of hand when pulling the head

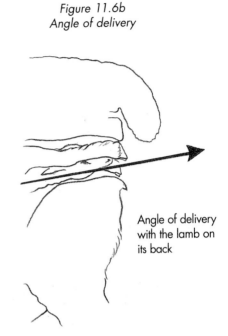

Figure 11.6a
Angle of delivery

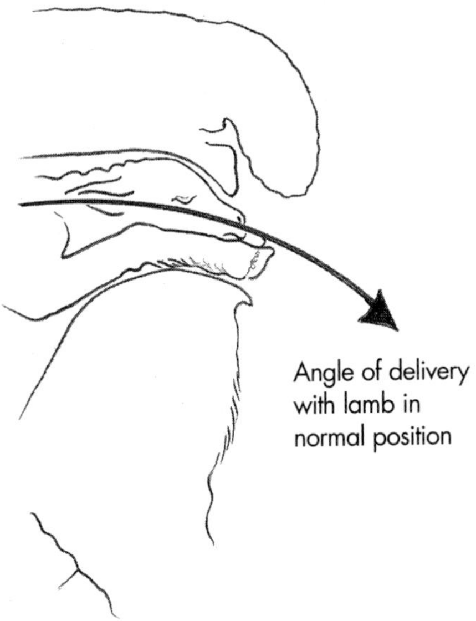

Angle of delivery
with lamb in
normal position

Figure 11.6b
Angle of delivery

Angle of delivery
with the lamb on
its back

Complications affecting the lamb

Usually, with the lamb's head and front feet lying in a streamlined position, the birth will progress without hindrance, but if the feet or nose are slightly out of line, the contractions of the ewe will gradually increase that deviation. So, if an apparently normal presentation is not progressing, it is important to check again after about half an hour so that any problem which has developed can be corrected before it becomes too bad. Problems can also arise where the lamb is abnormally large, even if it is presented correctly. Where the lamb is disproportionally large, a caesarean operation may be necessary. The size of the feet will usually give an idea of the size of the lamb.

Is the lamb alive or dead?

It is easy to tell a good vigorous live lamb, as it will pull back as you handle the legs, will suck a finger put into the mouth, or blink if the eyelid is touched. It is also easy to tell that there is a dead lamb when there is a foul smell and wool is coming away from the lamb. It is more difficult to tell whether a lamb is 'nearly dead' or 'just dead', and you sometimes get a pleasant surprise when a lamb which has shown no movements during delivery does come out alive and can be resuscitated! It is also important to know that occasionally there can be one very dead and smelly lamb followed by one or more live lambs.

CHAPTER 12

Ringwomb

Sometimes the neck of the womb does not dilate properly and this is known as ringwomb. This often follows prolapse of the cervix and is common with a breech presentation or a dead lamb, but may also occur for no apparent reason. Do not mistake a ewe which is in the early stages of lambing and in which the cervix is not yet fully dilated for a case of ringwomb – at this stage the rounded balloon-like end of the waterbag can be felt in the vagina or cervix; on no account rupture it, but give the ewe more time – at least a half to one hour – as she is not yet ready to lamb.

With a true case of ringwomb, the ewe has usually been uneasy for some time, the waterbag has appeared and ruptured but no more progress has been made. On examination, the cervix feels like a tight rubber band, sometimes allowing the entry of only the tip of one finger, sometimes several fingers (Figure 12.1). The neck of the uterus

Figure 12.1
Ringwomb – normal presentation

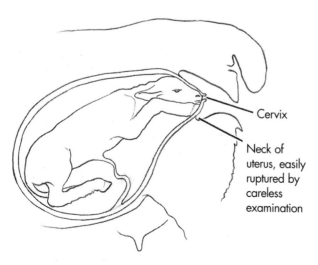

Cervix

Neck of uterus, easily ruptured by careless examination

Figure 12.2
Ringwomb – breech presentation

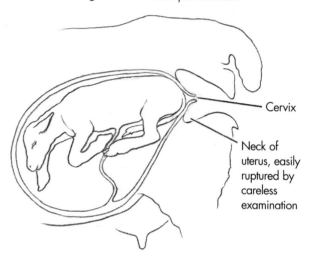

Cervix

Neck of uterus, easily ruptured by careless examination

is often stretched paper thin and parts of the lamb can easily be felt through it – take great care, for careless examination can easily rupture it. With a breech birth and some other malpresentations, the neck of the uterus is not tightly stretched and the entrance to the cervix will be 'floating' and may be difficult to locate at first (Figure 12.2).

Examine carefully with a well-lubricated hand. If membranes are present and you are certain that the ewe has been given plenty of time to lamb normally, it may be possible to dilate the cervix manually. It is very important that the ewe is handled quietly and calmly and great patience is needed. Serious injury can be caused by attempting to force a way into the uterus – tears of the cervix with severe bleeding and shock are common sequels where excessive force has been used.

Put the ewe on her side, lubricate the hand well, covering the fingernails with soap. Gently insert the tip of the finger into the cervix, repeatedly rotate the wrist and with gentle outward pressure massage the inside of the cervical ring. Lubrication must be continually re-applied; otherwise damage will occur. Gradually introduce another finger, until, with the wrist rotating all the time, all four fingers are being used. At this stage tuck the thumb and fingers into the palm of the hand, and using copious lubrication on the knuckles, continue massaging until access is gained to the uterus and an assessment of the position of the lamb can be made. Often, the cervix suddenly relaxes, although this can take some time, perhaps 20 to 30 minutes, so should only be attempted by those with patience, skill and gentleness!

If the cervix has opened, an examination must be made of the lamb to assess its position. It can then be delivered following the instructions for the appropriate presentation. However, even if the cervix does appear to have relaxed, it may not have opened to its full normal extent. If the lamb is a very large one, or is coming backwards, great care must be taken in manipulating and delivering the lamb, as there is a danger that the cervix may tear, causing severe bleeding or other damage.

If you are unsure about this procedure or are failing to make progress, consult your vet for it may be necessary to perform a caesarean operation.

CHAPTER 13

Visible Malpresentations and Problems

Head and one leg

This is the most common malpresentation, one front leg being stuck in either the uterus or the birth canal. In very rare cases the protruding leg can be the back leg of another lamb, so first check to see if it is front or back. If it is a case of two lambs coming together, see page 52.

If it is a front leg, lay the ewe flat on her back to check the position of the missing leg.

- If the missing leg is caught with the foot doubled up just inside the rim of the ewe's pelvis (Figure 13.1), the trapped leg must be pulled out (if it is not corrected before pulling, the leg will be doubled back compressing the lamb's chest). Crooking a finger around the first joint to straighten the leg, cup your hand over

the foot to protect the uterus. Then grasp the leg by the fetlock and pull it clear of the vulva. The ewe is then rolled onto her side and the lamb delivered normally by applying an equal pull on its head and both legs.

- If the missing foot cannot be felt as described above, the leg will be lying along the body of the lamb (Figure 13.2). If this problem has been spotted early enough and there is still plenty of natural lubrication left, the lamb can usually be delivered safely with the leg in this position, provided the lamb is not large (check the size of the feet) and the ewe has enough room in her pelvis. Try to decide whether this is possible before starting to pull, as it can be most difficult to correct a lamb which has been pulled with a

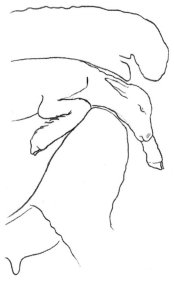

Figure 13.1
Head and one leg, other leg doubled back at knee

Figure 13.2
Head and one leg, other leg lying alongside body

leg back and which has then got stuck. Place the ewe on her side with the trapped leg uppermost; the lamb can then be 'sledged' out on its free leg by exerting an equal pull on the free leg and the head. Attempts to deliver the lamb with its trapped leg on the underside will probably result in the shoulder of the lamb becoming jammed against the pelvis.

- If the uterus is dry because the ewe has been straining for some time, it will be necessary to raise her hindquarters fully to allow copious lubrication to be inserted to flow along the body of the lamb. After this the ewe can be lowered, placed with the trapped leg uppermost and delivered as above.

- In some cases, particularly when dealing with a large lamb, it is not possible to deliver the lamb with the leg retained, so it must be brought forward. The ewe's hindquarters should be raised fully and copious lubrication inserted. Put a rope on the visible leg and push the leg slightly back in order to flex the elbow and shoulder joints. Then, with a well-lubricated hand (use the hand on the same side as the trapped leg), move the fingers along the trapped leg to the knee, crook a finger round the knee joint and draw forward until the leg is bent double. Then move the fingers to the foot and cup the foot with the fingers to protect the wall of the uterus. Keeping gentle pressure on the body of the lamb, bring the foot into line and pull the leg out fully. Then pull the other leg fully out. Lower the ewe onto her side and deliver the lamb by applying an equal pull on the head and both legs. The lamb may require immediate resuscitation (see page 64).

Head

Immediate action is necessary, as constriction of the neck of the lamb soon causes the tongue and head to swell, making the delivery more difficult and endangering the life of the lamb. You must **never** attempt to deliver the lamb with both legs retained. To do so will almost certainly mean seriously damaging, probably killing, the ewe. If the tongue or head is swollen when the lamb is delivered, see page 64 for resuscitation (the swelling will go in a few hours).

Place the ewe flat on her back to check the position of the feet.

- The front feet may be doubled up and stuck just inside the rim of the ewe's pelvis (Figure 13.3). If this is the case, they can usually be flipped out with the fingers. Keeping the ewe on her back, first draw out one leg fully and then the other one. Then put the ewe on her side and deliver the lamb by an equal pull on the head and both legs.

Figure 13.3
Head, both legs bent at knees

- If only one leg is stuck in this position, pull it out as described above and proceed as for 'Head and one leg' see page 38.

- More often the legs are lying right back along the body of the lamb, or even crossed under the chest (Figure 13.4). In this case it is essential to raise the ewe's hindquarters fully to allow the

Figure 13.4
Head, both legs lying alongside body

body of the lamb to drop slightly into the uterus, taking pressure off its chest and allowing more room for the hand of the shepherd. Copious lubrication is essential and should be squirted down each side of the lamb. Insert a well-lubricated hand with the palm facing the lamb, passing the hand carefully down the side of the head, neck and shoulder. Locate one leg, move the hand down until a finger can be crooked around the knee joint, draw the leg forward until it is bent double, then move the fingers to the foot. Protect the lining of the uterus by cupping the foot in the fingers, and whilst keeping gentle pressure on the body of the lamb, not the wall of the uterus, bring the foot into line. Then bring the foot forward **until the foot only** is just outside the vulva – no further at this stage, for if it is pulled right out, the space available for the shepherd's hand, which must be inserted again to retrieve the other leg, is decreased. Follow the same procedure for the other leg, but this one can be pulled out to its full extent. Then grasp the foot of the first leg above the fetlock and pull that fully out also. Lower the ewe's hindquarters to the ground,

put the ewe on her side and deliver the lamb by applying an equal pull on the head and both legs.

- If the legs cannot be located due to lack of space and the head is not grossly swollen, the head of the lamb must be returned to the uterus. Great care is needed, copious lubrication is essential and force must not be used, for this would risk rupturing the uterus leading to the death of the ewe. First, secure a head rope or snare in position on the lamb's head, then hoist the ewe's hindquarters fully, and with copious lubrication gently push the head back into the uterus, applying pressure only between the ewe's contractions. Throughout this manoeuvre, the head rope or snare must be kept under some tension to ensure the head remains in line. When the head is completely returned, whilst maintaining tension on the head rope or snare, locate one leg, and as described above bring **the foot only** outside the vulva and then fix a leg rope to that leg. Locate the other leg and in a similar manner bring **the foot only** outside the vulva and secure a foot rope to this leg also. At this stage the head of the lamb can be drawn into the vagina and out of the vulva by pulling on the head rope or snare, keeping slight tension on the leg ropes to ensure that they are not squeezed back inside. When the head of the lamb is clear of the vulva, wipe away any membranes from the nose and mouth of the lamb. Pull out first one leg fully, then the other. Lower the ewe, place her on her side and complete delivery by applying an equal pull to the head and both legs.

In rare cases, the head of the lamb may be too swollen to be returned to the uterus and it may not be possible to get a hand into the uterus to reach the legs. In this case the procedure depends on whether the lamb is alive or dead. To determine this, touch the corner of the eyes for a blink reflex, put a finger in the mouth to test for a suck reflex, or twist an ear.

- If the lamb is dead, the only option is to cut off the head to allow room for the shepherd's hand to locate the legs, which can then be straightened. Use a sharp knife to remove the head, cutting through the joint in the neck bones immediately behind the head. Be careful not to injure the ewe whilst doing this. Care should be taken to protect the stump of the lamb's neck with the fingers as the legs are retrieved and pulled to remove the body of the lamb. If this is not done, damage may be caused by the protruding sharp ends of bones.

- If the lamb is alive and it is not possible to get a hand into the uterus to reach the legs (in practice we have always found it possible eventually to get a small hand albeit at the cost of badly bruising it, into the uterus), the lamb must be humanely killed before carrying out the above procedure. Get veterinary help for this.

Nose and two legs

In this case the top of the lamb's head is jammed against the ewe's pelvis (Figure 13.5). It should be dealt with as soon as noticed, for, as the ewe strains and the thicker parts of the legs are expelled, the head

Figure 13.5
Nose and two legs

is squeezed back, leading to a more complicated 'head back' presentation; the legs must not therefore be pulled before the head is clear of the vulva.

The following procedure is best carried out with the ewe in the standing position, which lessens the pressure on the top of the lamb's head. First clear the lamb's nose and mouth of any membranes and check that the legs are front legs (see page 34). In rare situations one might be the back leg of another lamb (see 'Two lambs coming together', page 52). Having made sure this is not the case, firmly secure a leg rope on each leg, because as the head is drawn forwards the legs may be squeezed back into the uterus. Insert well-lubricated fingers along the side of the lamb's jaw and work them over the top of the head. Move the fingers to either side of the neck behind the ears and draw the head forward. If there is difficulty getting a grip, use a head rope or snare. Pull the head clear of the vulva, maintaining tension on the leg ropes to keep them in position. Next put the ewe flat on her back, relieving pressure on the lamb's elbows. Draw each leg out in turn, turn the ewe onto her side and complete delivery by applying an equal pull on the head and both legs.

Nose

- Check first for ringwomb (page 36). In this case the nose is trapped in the partially dilated cervix which will feel like a tightly stretched thin rubber band round the nose. If this situation is not immediately recognised and the hand is pushed up the side of the lamb's head, the cervix can be torn. The procedure for dealing with ringwomb should be followed (page 36).

- If this is not a case of ringwomb, in the majority of cases the top of the lamb's head will be jammed against the ewe's pelvis and the feet will be doubled up against the rim of the pelvis (Figure 13.6), as a result of the ewe's contractions. Do not attempt to pull the feet out at

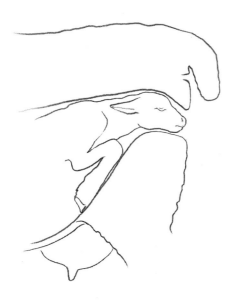

Figure 13.6
Nose

this stage, as such action is likely to result in the lamb's head being squeezed back, making the presentation far more complicated. The head must be brought out first and this is best achieved with the ewe in the standing position to relieve pressure on the top of the lamb's head. Clear the lamb's nose, lubricate the hand well, run the fingers along the side of the lamb's jaw and move up to the top of the head. Get the fingers over the top of the head, then either side of the neck behind the ears and pull the head out just clear of the vulva. Pressure must now be taken off the lamb's legs, so put the ewe flat on her back and, if the feet are just inside the rim of the pelvis, they can usually be flipped out with the fingers. Pull each leg out in turn to its full extent, roll the ewe onto her side and complete the delivery by applying an equal pull on the head and both legs.

- If only one leg can be pulled forward see 'Head and one leg', page 38.

- If no legs can be felt see 'Head', page 39.

Two legs

This could be either a lamb coming backwards, the head of the lamb turned back, or two lambs coming together.

Do not pull the legs until a careful examination has been made. With the ewe in a standing position, check the following:

- Are the legs front or back legs? See page 34.

- Do they belong to the same lamb? Check by running your fingers up the inside of one leg to the body of the lamb and continue down along the inside of the other leg. If they belong to separate lambs, see 'Two lambs coming together' page 52.

- If the legs are front legs, with the ewe still on her feet, insert a well-lubricated hand on to the top side of the lamb to see if the head is doubled back over its shoulder (Figure 13.7). If the head cannot be felt in this position, put the ewe flat on her back to see if the head is doubled back under the lamb's chest (Figure 13.8).

Figure 13.7
Front legs, head turned over shoulder

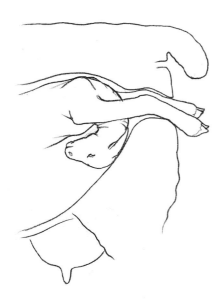

Figure 13.8
Front legs, head turned under chest

- If this situation has been spotted early enough, so there is plenty of natural lubrication present, and if the lamb is fairly small, the head can often be brought into the correct alignment by cupping your fingers over the nose of the lamb and gently bringing it into its natural position. **Never pull on the lower jaw** as this is likely to cause damage. Remember that the ewe must be flat on her back if the lamb's head is turned under its chest, or in a standing position if the lamb's head is turned back over its shoulder. When the head is correctly aligned, place the fingers over the back of the head, either side of the neck and behind the ears, and draw it out clear of the vulva. Wipe the nose and mouth clear of membranes, then put the ewe on her side and deliver the lamb by applying an equal pull on the head and both legs.

- If the legs are front legs and the head cannot be felt, or the head cannot be aligned, the legs will have to be returned into the uterus before any attempt can be made to locate and secure the head of the lamb. The hindquarters of the ewe

must be raised about halfway, just sufficient to assist in the following procedure, but not so far as to cause the lamb's head to fall further into the uterus. Secure a leg rope onto one leg and, with copious lubrication, gently push the leg back into the uterus so as to flex the elbow against the lamb's chest, applying pressure between the ewe's contractions. Do the same with the other leg. In some cases it may be necessary to fold the legs beneath the lamb, but if this is done, they should always be straightened as soon as the head has been brought back into line. Remember to note which rope belongs to which leg, so there is no danger of them becoming crossed during delivery. There will now be much more room to locate the head of the lamb; it may be in the correct alignment with its body, or lying back over its shoulder, or tucked under its chest. It should be possible, with the aid of copious lubrication, to bring the head into the correct alignment by cupping the fingers under the chin (not pulling on the jaw), although it can sometimes be difficult to fully straighten the head without the aid of a head rope or snare. If, whilst you are attempting to straighten the head, it tries to return to its original position, it is probable that the head is being rotated in the wrong direction. When the head is in the correct alignment, secure a head rope or snare in position if you have not already applied one, and maintain tension to prevent it becoming displaced again. Check that the legs are correctly aligned. Raise the ewe's hindquarters higher and pull the head out clear of the vulva, clean it of membranes, and then pull out each leg in turn to its full extent. Lower the ewe onto her side and deliver the lamb by applying an equal pull on the head and both legs.

- If the legs are back legs (Figure 13.9), the problem in most cases will be a large lamb in a small ewe which has been straining for some time; most of the fluids will have been lost,

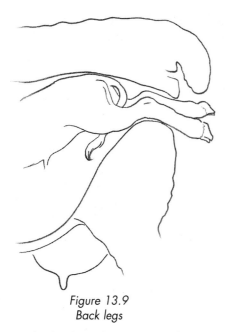

Figure 13.9
Back legs

leaving a dry lamb in the uterus. Before starting the following procedure, make sure that there is sufficient room to pull out the lamb quickly, ie. you are not too close to a wall or hurdle and that the necessary equipment to resuscitate the lamb is to hand, since large lambs coming backwards are at great risk of suffocation because the umbilical cord is compressed or separates during delivery. Copious lubrication must be used. If the legs are not fully out of the vulva, secure a leg rope to each and maintain a gentle tension to prevent them sliding back whilst hoisting the ewe's hindquarters fully. Insert plenty of lubrication down the sides of the lamb by moving the soaped hand along its body. Then insert more lubricant and pull out each leg in turn to its full extent to the point where the root of the tail is just in the birth canal. If the legs do not come out easily they may be locked against the brim of the pelvis (Figure 13.10), so pass the hand under each of the lamb's stifle joints in turn and lift gently to release. Lubricate again, then lower the ewe onto her side, and with an equal, slightly rotating pull on both legs and then the hips, deliver the lamb as quickly as possible. If the lamb is very large, it can be rotated by 45° by

crossing the back legs before pulling – this ensures the maximum amount of space for the chest of the lamb as it passes through the ewe's pelvis. Immediately clear away any fluid and membranes from the nose and mouth of the lamb, then hold it upside down by the back legs for a few seconds, wiping away any further mucus from the nose. Then lay the lamb down and apply further resuscitation if necessary (see page 64).

- If the back legs are fully out of the vulva, leg ropes are not necessary, but the same procedure should be followed to ensure a quick delivery of the lamb.

There are possible hazards for both the ewe and lamb with a large lamb in posterior presentation:

- the lamb may suffocate

- the lamb may sustain broken ribs which is often fatal

- the lamb may suffer a ruptured liver which is fatal

- the ewe may have a higher risk of uterine prolapse.

In spite of these risks, you should not attempt to turn the lamb. This is not possible with a large lamb and any attempt would be likely to result in damage to the ewe, such as tearing of the uterus. If a particularly large valuable lamb is found in posterior presentation, your vet should be consulted immediately about the advisability of carrying out a caesarean operation.

One leg

This could be the lamb coming backwards with one leg out and one retained (a 'half breech'), or it may be a lamb coming forwards with a front leg out and the head and other leg either jammed at the ewe's pelvis or doubled back. First establish if the leg is a front or back leg (see page 34).

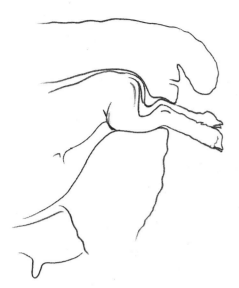

Figure 13.10
Back legs, stifles locked

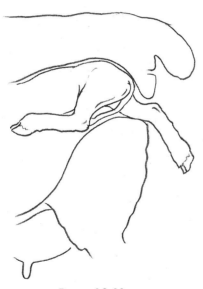

Figure 13.11
One back leg

* If it is a back leg (Figure 13.11), the other back leg must be brought out before delivery is possible. Before starting the following procedure, make sure that there is sufficient space to pull out the lamb quickly, ie. that you are not

too close to a wall or hurdle, and that the necessary equipment to resuscitate the lamb is to hand, since large lambs coming backwards are at great risk of suffocation because the umbilical cord is compressed or separates during delivery. Raise the hindquarters of the ewe and insert copious lubrication down the sides of the lamb. Using the appropriate hand, ie. the same side as the missing leg, move gently along the leg until the fingers reach the foot. Cup the fingers over the foot to avoid damage to the wall of the uterus and place the heel of the hand onto the hock of the lamb's leg, pressing it gently into the body of the lamb. With the fingers, move the foot in a natural arc (by turning the toe slightly towards the midline) into correct alignment with the body of the lamb, then pull the leg out level with the other leg. Pull both legs out in turn to their full extent, to the point where the root of the tail is just in the birth canal – no further at this stage. Now lower the ewe, put her on her side and with an equal, slightly rotating pull on both legs and then the hips, deliver the lamb as quickly as possible. Immediately clear away any fluid and membranes from the nose and mouth of the lamb, then hold it upside down by the back legs for a few seconds, wiping away any further mucus from the nose. Lay the lamb down and apply further resuscitation if necessary (see page 64). For the hazards associated with a large lamb coming backwards see page 44.

* If the leg is a front leg, with the ewe in a standing position, insert a well-lubricated hand on the top side of the lamb to see if the head is doubled back over its shoulder (Figure 13.12) If the head cannot be felt in this position, put the ewe flat on her back to see if the head is doubled under the lamb's chest (Figure 13.13). If this situation has been spotted early enough, so there is plenty of natural lubrication present, and if the lamb is fairly small, the head can

Figure 13.12
One front leg, head turned over shoulder

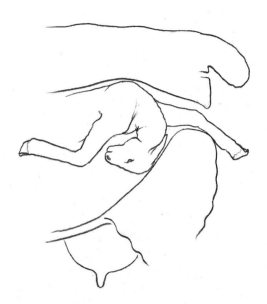

Figure 13.13
One front leg, head turned under chest

often be brought into the correct alignment by putting the thumb below the eye and pushing, at the same time cupping the fingers over the nose of the lamb and gently bringing it into its natural position. **Never pull on the lower jaw** as this is likely to cause damage. Remember that the ewe must be flat on her back if the lamb's head is turned under its chest, or in a standing position if the lamb's head is turned back over its shoulder. When the head is correctly aligned, place the fingers over the back of the head, either side of the neck and behind the ears, and draw it out clear of the vulva.

- Check whether the trapped leg is doubled up just inside the rim of the ewe's pelvis. If this is the case, put the ewe flat on her back, and with a finger round the first joint, flip out the foot, grasp the leg above the fetlock and draw the leg out to its full extent. Then put the ewe on her side and deliver the lamb by applying an equal pull on the head and both legs.

- If the trapped leg cannot be felt it will be lying down the side of the lamb. Providing the lamb

is not a large one, lay the ewe on her side with the protruding leg of the lamb nearest the ground, thereby taking pressure off the leg which is trapped. Then with an equal pull on the head and free leg, complete the delivery.

- If the head cannot be located, or there is not enough room to correct the position of the head as described above, the protruding leg will have to be returned to the uterus. The hindquarters of the ewe should be raised about halfway, just sufficient to assist in the following procedure but not so far as to cause the lamb's head to fall further into the uterus. First check if the trapped leg is doubled up just inside the rim of the ewe's pelvis. If this is the case, do not pull the leg out at this stage. Straighten out the foot, secure a leg rope in position, and using copious lubrication, gently push the leg back into the uterus, applying pressure between the ewe's contractions. Do the same with the free leg, remembering to note which rope belongs to which leg, so that there is no danger of them being crossed during delivery. It should be possible, with the aid of copious lubrication, to bring the head into correct alignment by cupping the fingers under the chin (not pulling on the jaw), although it can sometimes be difficult to fully straighten the head without the aid of a head rope or snare. If whilst you are attempting to straighten the head, it tries to return to its original position, it is probable that the head is being rotated in the wrong direction. When the head is in the correct alignment, secure a head rope or snare in position if you have not already applied one, and maintain tension to prevent it becoming displaced again. Check that the legs are correctly aligned. Raise the ewe's hindquarters higher and pull the head out clear of the vulva, clean it of membranes, and then pull out each leg in turn to its full extent. Lower the ewe onto her side and deliver the lamb by applying an equal pull on the head and both legs.

- If the trapped leg was not located just inside the rim of the ewe's pelvis, it will be lying down the side of the lamb's body and, providing the lamb is not a large one, can be left in that position during delivery. Using copious lubrication follow the procedure just described. When the head and free leg of the lamb are in place outside the vulva, lower the ewe, put her on her side with the trapped leg uppermost to relieve the pressure on that leg, and deliver the lamb by applying an equal pull on the head and free leg.

- If the lamb is large, the trapped leg should be brought forwards, with a well-lubricated hand (use the hand on the same side as the trapped leg). Move your fingers along the trapped leg to the knee, crook a finger round the knee joint and draw forwards until the leg is bent double. Then move your fingers to the foot and cup the foot with the fingers to protect the wall of the uterus, and keeping gentle pressure on the body of the lamb, bring the foot into line. Put on a leg rope, and with the head in the correct alignment and maintaining tension on the head

rope or snare, raise the ewe's hindquarters higher and pull the head out clear of the vulva. Clean it of membranes, then pull out each leg in turn to its full extent. Lower the ewe onto her side and deliver the lamb by applying an equal pull on the head and both legs.

CHAPTER 14

Internal Malpresentations and Problems

Membranes

In a normal lambing the waterbag, which is thin and more or less colourless, is expelled and bursts shortly before the lamb is born. If the waterbag has burst but there has been no further progress after about an hour, or if thick fleshy membranes are hanging from the vulva, this may signify that something is wrong.

* Note the colour of the membranes and fluids.

* Check for any unpleasant smell.

* Check for ringwomb (see page 36).

If the membranes are a brownish colour and smell unpleasant, it is likely that a dead lamb is present (see 'Dead smelly lambs' page 54).

If the membranes are a healthy colour, there is no unpleasant smell and the cervix is fully dilated, examine the ewe in the standing position with a well-lubricated hand. If the problem is not obvious, try a further examination with the ewe on her side, and then flat on her back. If no conclusions can be drawn from these examinations, raise the ewe's hindquarters slightly (no further than absolutely necessary) and then try again to establish the lamb's position. If you are unsure, consult your vet.

Occasionally it is possible to be caught out by a ewe which has lambed unobserved, and whose lamb has been 'stolen' by another ewe near lambing. If no lamb can be felt by examining the uterus, check by ballotting externally (see pages 56-7). If no lamb is felt, check for the thief!

Head doubled back

The head of the lamb may either be doubled back over its shoulder (Figure 14.1) or doubled up under its chest (Figure 14.2a).

Figure 14.1
Head doubled back, head over shoulder

When the head of the lamb is doubled up under its chest and the legs are back, this presentation feels very similar to a breech birth, especially if thick membranes are present (Figure 14.2b). In order to quickly make an assessment in this case, it is necessary to find either 'teeth or tail'. Slide a well-lubricated hand along what is believed to be the head; if teeth are felt, the presentation is confirmed. If a tail is discovered (and that is not always as obvious as one would imagine, sometimes being tightly folded under the body in thick membranes, or twisted up under the ewe's pelvis), this is a breech presentation (see page 49).

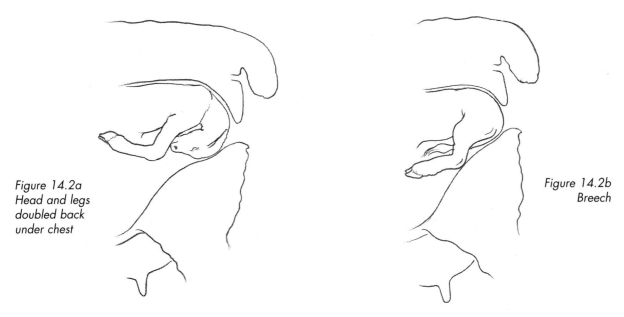

*Figure 14.2a
Head and legs
doubled back
under chest*

*Figure 14.2b
Breech*

*Figures 14.2a–b
These two presentations can feel very alike and are easily confused*

- To correct the position of the head, the hind-quarters of the ewe must be raised slightly – no further at this stage or the head will slip back into the uterus. Copious lubrication must be used.

- If the head is folded back over the lamb's shoulder, it must be gently brought into the correct alignment, but care must be taken not to rotate it in the wrong direction. If the head tries to return to its original position, this indicates that an error may be being made. Occasionally it is necessary to attach a head rope or snare to assist in realignment.

- If the head is folded under the lamb's chest, this is usually easily corrected by tucking the fingers under the chin (not holding the jaw) and bringing the head into line.

Once the head is correctly aligned, fix a head rope or snare in position, unless already in place. Whilst keeping tension on the head rope or snare at all times, locate the legs and bring them into line, fix a rope on each leg, noting which rope belongs to which leg so that there is no danger of crossing them during delivery. **Do not draw the legs forward at this stage.** Raise the ewe's hind-quarters higher, then draw forwards the lamb's head until the **nose only** is just out of the vulva. Clear away any membranes from the nose and mouth of the lamb, then draw the legs forwards until each foot is lying at either side of the lamb's head. Draw the head out fully, clear of the vulva, check again that the nose and mouth are clear of membranes, and then draw out each leg in turn to its full extent. Lower the ewe, put her on her side and deliver the lamb by an equal pull on its head and both legs.

Breech

In a breech presentation the back legs are lying under the abdomen of the lamb pointing towards the shoulders. How far depends on the length of time the ewe has been lambing. Ewes with this presentation are often missed in the early stages, as they may not show much sign of straining and the waterbag may rupture within the birth canal. The only clues may be uneasiness and wetness down the skin beneath the vulva, particularly over the back of the udder. Any ewe showing this picture should be carefully examined, since the earlier this presentation is found, the easier it is to correct. If the examination shows a normal presentation, the ewe can be safely allowed more time. If this malpresentation is not spotted early, the contractions of the uterus gradually push the hind end of the lamb into the pelvis, the legs are forced further in until they point towards the shoulders of the lamb, and the fluids are lost making correction difficult. A breech birth is usually an indication of a lack of room in the uterus for the lamb to extend its legs – either a large single or the first of triplets.

As mentioned previously, it is very easy to confuse this presentation with one where the lamb's head is doubled under the chest and the legs are back (Figure 14.2a). Therefore on first examination seek for either 'teeth or tail'. Discovering the teeth will confirm that the head is doubled under the chest (see page 49). The tail may be mistaken for an ear, or is often tightly folded under the body of the lamb; once the tail has been identified, this is confirmed as a case of breech birth (Figure 14.2b).

Before the lamb can be delivered, both feet must be straightened and brought clear of the vulva. This can only be safely achieved by raising the hindquarters of the ewe, thus enabling the lamb to drop back into the uterus and allowing room for the repositioning of the legs.

Before starting the following procedure, make sure that there is sufficient room to pull out the lamb quickly, ie. you are not too close to a wall or hurdle, and that the necessary equipment to resuscitate the lamb is to hand, since large lambs coming backwards are at great risk of suffocation because the umbilical cord is compressed or

separates during delivery. There is also a great risk of the uterus being damaged or ruptured, so the utmost care is needed whilst moving the legs of the lamb.

* Raise the hindquarters of the ewe about halfway. With copious lubrication and using the appropriate hand, ie. the same side as the missing leg, insert the hand under and alongside one leg of the lamb until the fingers reach the foot. Cup the fingers over the foot to avoid damage to the wall of the uterus, press the heel of the hand onto the hock of the lamb's leg, pressing it gently into the body of the lamb. Then with the fingers move the foot in a natural arc (by turning the foot slightly towards the midline) until it is lying in line with the lamb's body and the foot is in the birth canal – no further at this point. With the other hand, repeat the manoeuvre with the other foot and then pull each leg out to the hock, but no further. Now lower the ewe, put her on her side and with an equal, slightly rotating pull on both legs and then the hips, deliver the lamb as quickly as possible. Immediately clear away any fluid and membranes from the nose and mouth of the lamb and hold it upside down by the back legs for a few seconds, wiping away any further mucus from the nose. Then lay the lamb down and apply further resuscitation if necessary (see page 64). For the hazards associated with a large lamb coming backwards see page 44.

* If you are unable to correct the position because the uterus is so tightly clamped down that there is a severe danger of rupturing the uterus, get help from your vet.

Back

Very occasionally a lamb will be presented with its back towards the birth canal. This situation probably arises as an extreme result of a head doubled under the chest, or a breech presentation, when the ewe has not been noticed and uterine

Figure 14.3
Back, ribs

Figure 14.4
Back, loin

contractions force the body of the lamb into this abnormal presentation. Unless you have had plenty of experience, it is better to let the vet tackle this one, as it can be very difficult to correct without injuring the ewe. If you feel you can proceed yourself, carry out a careful examination, preferably with the ewe in the standing position and with the aid of copious lubrication, trying to find a landmark on the lamb. This will usually be either the ribs (Figure 14.3) or the loin (Figure 14.4). If the ribs are felt, raise the hindquarters of the ewe slightly and try to reach the head, then treat as a head doubled under the chest (see page 49). If the loin is felt, raise the hindquarters of the ewe about halfway and gradually convert to a breech, then follow the instructions for that presentation (see page 49). In either case, copious lubrication, great patience and care are needed; otherwise the uterus may be ruptured.

Large single lamb

Sometimes a large single lamb which is correctly presented may cause problems in a ewe which has a small pelvis, eg. a first-time lamber, particularly a Texel or Beltex, or a small hill ewe. It is

particularly important not to interfere too early in these cases, allowing plenty of time for the contractions of the ewe to push the lamb's head into the pelvis. If the legs are pulled before the head is in the pelvis a difficult 'head back' situation will be created. However, if no progress is made in spite of having allowed plenty of time, the ewe can be assisted most easily as follows:

- With the ewe's hindquarters slightly raised, check that the head is correctly aligned, then put a head rope or snare in position on the lamb. Keep a slight tension on the rope with the other hand, but do not pull at this stage. Then place a rope on each leg in turn, being careful that each rope is positioned correctly above the fetlock and leading off from underneath the leg (Figure 4.5 page 10). Check that the legs are correctly aligned, with the feet just at the rim of the ewe's pelvis. Keep sufficient tension on the ropes to prevent the legs from falling back, but do not pull them at this stage. Then use the head rope or snare to pull the lamb's head into the pelvis, until the nose is just clear of the vulva. Now pull each leg rope in turn until the feet are lying either side of the nose. The head can then be drawn fully out, clear of the vulva. Check that the nose and mouth are clear of membranes, then draw out each leg in turn to its full extent. Lower the ewe, put her on her side and deliver the lamb by an equal pull on its head and both legs.

Figure 14.5
Two lambs coming together forwards

- If the lamb's head is too large to enter the ewe's pelvis, it is likely that a caesarean operation will be necessary so consult your vet.

Two lambs coming together

Occasionally, parts of two lambs are presented together in the birth canal. This creates a jam which must be sorted out before either lamb can be delivered. In this case a decision has to be made about which lamb is easiest to deliver first, and this will usually be the one nearest the vulva. If both lambs remain within the uterus and it is not obvious which should be delivered first, see page 53.

The possible ways in which the lambs are likely to be presented are:

- both coming forwards (Figure 14.5)
- one coming forwards and one backwards (Figure 14.6)
- both coming backwards (rare) (Figure 14.7).

The exact situation must be established by carrying out a careful internal examination with the aid of copious lubrication.

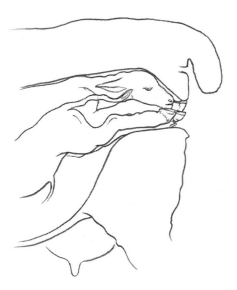

Figure 14.6
Two lambs coming together, one backwards and one forwards

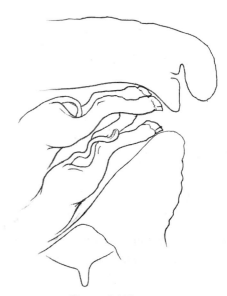

Figure 14.7
Two lambs coming together backwards

- If, on initial examination, a lamb's head is located, secure a head rope or snare in position before proceeding any further. Keeping tension on the head rope or snare to prevent the lamb sliding back into the uterus, raise the hindquarters of the ewe just sufficiently to sort out the legs of the lamb. First locate a front leg by feeling down the neck of the lamb to the shoulder, then the knee, then the foot. Place a rope on that leg. Follow the same procedure using the other hand to locate and rope the other front leg. Note which rope belongs to which leg so that they do not become crossed, and make sure that the legs are lying in line but do not pull the legs at this stage. Check again that the head is also lying correctly aligned.

Up to this point, make no attempt to pull the lamb, because the other lamb lying at its side is creating too much pressure. The situation can be resolved as follows:

- With copious lubrication, make sure the head and leg ropes are lying in line with the lamb to the outside of the vulva. Keeping equal tension on these three ropes all the time, raise the hindquarters of the ewe fully so that only her shoulders are touching the ground (this position must be held for no longer than absolutely necessary). This allows the second lamb to fall back down the uterus making room for the first lamb to be delivered. With the aid of the head rope or snare, draw the head clear of the vulva. Quickly clean the nose of the lamb, and pull out first one leg then the other. With equal tension on all ropes, draw the lamb out with a steady pull. As soon as this lamb has been delivered, lower the ewe immediately, put her on her side and place the lamb at her nose. After a few moments' rest, get her up on her feet. Do not attempt to deliver the second lamb at this stage.

- If, on initial examination, a lamb's head cannot be located or is in an abnormal position, raise the ewe's hindquarters slightly, and with a well-lubricated hand, examine the legs to see whether they are front or back ones (see page 34). If they are all front legs, a head will need to be located and, with great care, brought into line, before proceeding as above.

- If there is a mixture of front and back legs, it is often easiest to deliver the lamb which is coming backwards first. (The only time this may cause a problem is if the ewe's birth canal is tight or the lamb is very large, as it may get stuck and suffocate or sustain broken ribs during delivery – in this situation it is better to try to get the lamb coming forwards first if this is possible without damaging the ewe). Find two back legs which belong to the same lamb, by sliding the fingers along the inside of one leg to the body of the lamb and continue down the inside of the other. Fix a leg rope in position on each and raise the ewe's hindquarters a little higher, inserting copious lubrication down the sides of the lamb. Then, maintaining tension on both ropes, raise the hindquarters of the ewe fully, allowing the second lamb to slide down into the uterus. Pull out each leg in turn as far as the hock, then using an equal pull on both legs, draw the lamb steadily out to the point where the root of the tail just appears at the vulva. Give a quick decisive pull to deliver the lamb immediately as the cord will become compressed or separate and the lamb will be in danger of suffocation. As soon as the lamb is clear of the vulva, clean its nose and mouth and hold upside down by its back legs for a few seconds; squeeze its nose and wipe away any further fluid. Put the lamb down, immediately lower the ewe and put her on her side with the lamb at her nose. After a few moments, get her to her feet. Do not attempt to deliver the second lamb at this stage. If the lamb needs further resuscitation see page 64.

- After the first lamb has been delivered and the ewe is on her feet, leave her in peace for a while to recover, but keep her under close observation. If the second lamb is known to be malpresented, give appropriate assistance after the ewe has had a rest. If the second lamb is straight but has not been born after about half an hour, carry out a further examination and assist as necessary. Be careful to allow the ewe's own contractions to push the lamb towards the birth canal rather than pulling the lamb from deep within the uterus; otherwise she may prolapse the uterus. After the second lamb has been born, remember to make sure that no other lambs remain.

- If the ewe has given up any effort to lamb but both lambs remain within the uterus, neither having been pushed into the birth canal, a decision must be made on which lamb to deliver first. Copious lubrication is necessary as most of the birth fluids are likely to have been lost. Check what parts of the lambs can be felt, and which direction the lambs are presented. If hind legs are felt, it is often easiest to deliver this lamb first, as there will then be more room to locate the head of the second lamb. The only time this may cause a problem is if the ewe's birth canal is tight or the lamb is very large, as it may get stuck and suffocate or sustain broken ribs during delivery – in this situation it is better to try to get the lamb coming forwards first if this is possible without damaging the ewe. Great care must be taken that the lambs are not pulled too quickly from deep within the uterus, for this may lead to the ewe prolapsing the uterus. Try to allow the ewe's own contractions to assist in moving the lambs towards the birth canal.

Dead smelly lambs

The first indication that the lamb is dead is usually the smell. Within a few hours of the lamb dying there will be an unpleasant sweetish smell, detectable when the hand has been inserted to carry out an internal examination. Later the smell becomes increasingly putrid. The lamb will feel dry to the touch, its wool may have started to come away and it may be blown up with gas (a crackly feeling on pressing the skin).

Before attempting to deliver the lamb, have a bucket or plastic sack to hand so that the dead lamb(s) can be removed immediately and not contaminate the lambing area. Dead lambs must be delivered using all the principles required for a live lamb, but it is particularly important that the lining of the uterus is not damaged as the ewe may become seriously ill. Copious lubrication is always required to replace the lost natural fluids. If you are unable to deliver the lamb consult your vet, as it is possible that an embryotomy (dismembering) is necessary.

Always check for the presence of further lambs. It is not uncommon for a healthy live lamb to be still in the uterus at the same time as a dead smelly one, as the membranes effectively separate the individual lambs. If a live lamb is felt, check if it is in an abnormal position and correct as necessary. Try to allow the contractions of the ewe to move the lamb towards the birth canal, rather than pulling it from deep within the uterus, as doing this can risk causing the uterus to prolapse.

Occasionally, if the ewe has been trying to lamb unnoticed for several days and the lambs have started to putrefy, the lamb may come away in several pieces. Be sure to use copious lubrication and take great care in any manipulations. If the placenta is loose remove it carefully, but do **not** pull if still attached in the uterus.

An injection of antibiotic will be necessary after removal of smelly lambs, and it may be worth asking your vet for advice on other treatment, such as an NSAID (nonsteroidal anti-inflammatory drug), to help guard against the toxaemic effects on the ewe.

Deformed lambs

Deformed lambs are not uncommon, with a variety of problems including joints which will not bend correctly, duplication of limbs, conjoined twins, accumulation of fluid under the skin or in the abdomen. Another alarming deformity is called a schistosome or 'inside-out' lamb, in which the abdomen has not formed properly so that the lamb's intestines are exposed (Are they the lamb's or the ewe's intestines? – Get the vet!).

Attempting to lamb any of these can be extremely confusing. Any examination should be carried out carefully with copious lubrication, trying to assess whether it is possible to remove the lamb without damaging the ewe. If you are uncertain, consult your vet who may have to dismember the lamb (embryotomy) or carry out a caesarean operation. Most badly deformed lambs are dead or die soon after birth.

Torsion of the uterus

Rarely, a twist of the birth canal may develop just before lambing is due. The ewe is uneasy, but shows no progress and should always be examined – if she is not ready to lamb or there is no problem, more time can be allowed. The twist usually takes place in the region of the cervix and anterior vagina, although it may also occur anterior to the cervix, in which case diagnosis is difficult or impossible. The tightness and position of the twist determine what can be felt when an internal examination is carried out. If the twist is not too severe, it may be possible to reach the lamb, but the twisting direction of the birth canal is felt as the hand is inserted. More often, the twist is so tight that the hand feels just a tight spiral, like putting the hand into the twisted sleeve of a coat.

It is not safe to try to correct the twist by internal manipulation. If the twist is only slight and the lamb is correctly presented, raising the hindquarters of the ewe with the hoist may allow the torsion to correct itself. However, often the cervix is not fully dilated, so the case has then to be treated as a ringwomb (see page 36).

Generally the only way to deal with these cases, other than by a caesarean, is by 'unrolling' the ewe around the uterus. Do not attempt this procedure unless you are confident you will not damage the ewe.

* Two people are needed. Carefully place the ewe on her side. Try to assess, by feeling, the direction of the twist so that the ewe can be rolled in the appropriate direction, although this is easier said than done! Whilst one person keeps a hand in the vagina, the ewe should be carefully rolled over onto her other side. If this manoeuvre results in tightening the twist, the ewe must immediately be rolled in the other direction. The rolling over in the one appropriate direction should continue only until the twist has disappeared – this will usually require rolling the ewe at least 360°, perhaps more – only by feel will you be able to tell how far is necessary. When the twist has disappeared, feel carefully whether the cervix has dilated fully and for the presentation of the lamb. It is common for the cervix to be only partially dilated, so allow the ewe another half hour or so. If the cervix does not dilate, treat as for a case of ringwomb (see page 36). If at any time you are uncertain how to proceed, consult your vet.

CHAPTER 15

The Ewe After Lambing

In a normal, unaided lambing, the ewe will get to her feet as soon as the lamb is born and will turn to the lamb and begin licking it, all the time 'talking' to it with a characteristic soft bleating. This action:

- clears the lamb's nose of membranes

- stimulates breathing

- begins the mutual recognition process

- allows the uterus to fall back into the abdomen into its correct position.

If there are more lambs to be born, the interval between lambs is variable, some ewes producing them within a few minutes of each other. Most ewes spend 10 to 15 minutes licking the first lamb and sometimes allow it to suck before getting on with producing the next one. Problems which can arise at this time are:

- the first lamb sucks most of the colostrum if the ewe has only enough for one lamb

- the lamb may wander away whilst the ewe is producing the next one

- the lamb may be 'stolen' by another ewe which has not yet had her lamb.

If ewes are lambing outdoors with plenty of space, particularly if most are having singles, all that is required is careful observation to ensure that all is well. If ewes are lambing indoors, or outdoors in a crowded field with many producing twins, moving the ewe and lamb(s) to an individual lambing pen close by will prevent lambs getting lost or stolen. The bonding of ewe and lamb(s) is very important for the survival of the

lambs. Some ewes are better mothers than others, and once a bond has broken it is very difficult to get it re-established. It is good practice to leave the ewe and lambs in an individual pen for 24 hours in the case of a multiple birth, longer if the ewe is an inexperienced mother, or if there is any doubt about the health of the ewe or lambs.

Lambing pens should be bedded with clean dry straw and cleaned out between ewes if at all possible. A good supply of clean water is essential for the ewe in order for her to produce plenty of milk, and forage, preferably good-quality hay, should be provided in a rack. Feeding silage, particularly if it is low in dry matter, can make individual pens wet, producing conditions where navel or joint ill may become problems.

Checking for more lambs

After assisting a ewe to lamb, a check should always be made for the presence of another lamb in the uterus. This can be done in two ways, (1) by feeling the abdomen externally, or (2) by feeling inside the uterus with the hand.

1. If it has not been necessary to insert the hand into the uterus so far, then it is better to check, if possible, by feeling externally (this is called ballotting). If the ewe is lying down, the hand, made into a fist, should be pushed gently but firmly into the abdomen just in front of the udder (Figure 15.1). If there are one or more lambs still to be born it is usually possible to feel the fist knock against the body of the lamb. Alternatively, if the ewe is standing, circle the abdomen from behind with both arms, again in front of the udder, and lift gently up and down (Figure 15.2). The abdomen of a ewe which has no more lambs to come is generally relaxed and easy to lift. Those which still have lambs inside are

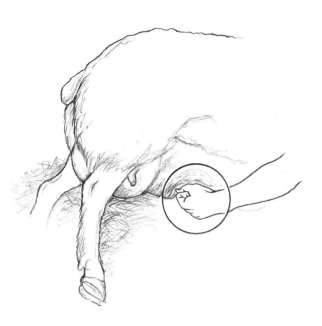

Figure 15.1
Ballotting the abdomen to feel for more lambs

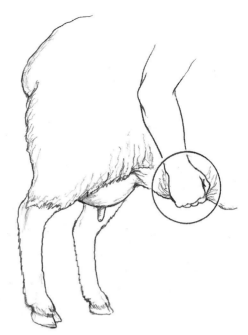

Figure 15.2
Lifting the abdomen to feel for more lambs

heavier to lift and the body of the lamb can generally be felt. Practice this technique on ewes which are just about to lamb or have just finished lambing, for it is an easy and safe method of checking any ewe to find out whether lambing is complete.

2. If it has already been necessary to enter the uterus to deliver one lamb, make sure your hand is clean and lubricated. Then, preferably with the ewe standing, carefully insert the hand into the uterus and feel around, remembering that the uterus of the sheep has two horns in which lambs can be lying. After a gentle examination, withdraw the hand slowly and carefully so as not to stimulate excessive straining.

* If there is another lamb in the uterus, the ewe must be kept under close observation to make sure that delivery proceeds normally.

* If the lamb is felt far away in the uterus, it should not be pulled as this may lead to inversion or prolapse of the uterus; the ewe should be left to move the lamb by means of her own contractions providing it is not

malpresented. If she has not delivered the lamb after about half an hour, she must be re-examined to see if the lamb has become malpresented, in which case this should be corrected and the lamb carefully delivered.

* If the ewe is too exhausted to strain, then she should be helped immediately. In this case, put her on her side and move a well-lubricated hand around the lamb. This should stimulate contractions pushing the lamb towards the birth canal. As the lamb enters the pelvis, it can be helped by gentle traction.

* Always check for more lambs, even after the second or third. The largest number of lambs delivered by one of us from a single ewe is six!

Routine postlambing ewe care

Getting the ewe to her feet

After an assisted lambing, the ewe should be encouraged to stand straightaway, following the pattern of a normal unaided lambing.

If the ewe is not able to stand because of exhaustion or illness, or is not released from any restraint immediately, she may begin to strain again precipitating a prolapse of the uterus. This can be a particular danger if the ewe is left lying on sloping ground with the hips lower than the shoulders.

If she will not stand, she should be supported in a half-standing position on her back feet only, for a few minutes, and then placed on her breastbone on level ground, with the lamb at her head, to encourage her to lick the lamb and start the bonding process.

Antibiotic treatment

You will need to take advice from your vet about antibiotic treatment for ewes which have been assisted to lamb. In general, an injection of long-acting antibiotic is advisable where manipulations have to be carried out within the uterus, but **this is not a substitute for good hygiene!**

Checking the udder

Before you leave the ewe and lambs, the udder must be checked to make sure there is a good supply of colostrum in both teats. Take care not to damage the teats by handling roughly – use of the birth fluids of the lamb or from the ewe's vulva to lubricate the fingers may be helpful. Draw each teat in turn – this removes the waxy plug from the end of the teat making it easier for the lamb to suck, as well as making sure colostrum is present.

If colostrum cannot be drawn from the teats, this may indicate that the lambs will need extra colostrum supplying (although lambs often appear to be able to get colostrum when we cannot!). If both sides of the udder appear empty, it may be because the sheep is old or in poor condition – colostrum and milk will probably be produced after a delay of a few hours, but the lambs must be supplied from another source in the meantime (see page 68). If colostrum can only be drawn from one teat and the other feels either empty or thickened, this indicates that the ewe has had mastitis sometime in the past, probably following the previous weaning. Milk is unlikely to be produced in this side of the udder, so she should preferably be left with only one lamb to rear. If foster mothers cannot be found for the surplus lambs, they can be left with the ewe, but one or more lambs will need to be supplemented with a bottle, probably until they are old enough to eat creep feed or grass.

The placenta

After the ewe has licked and fed the lambs, she usually lies down and the placenta is expelled, generally within a couple of hours after the last lamb has been born. The placenta should be removed and disposed of (by burying or burning); if left, it may be a source of infection, and outdoors will attract predators. Bear in mind that some ewes eat the placenta immediately it has been produced, although it is better to remove it if possible.

Retained placenta

This may indicate that:

- the ewe has lambed early or aborted

- there is another lamb still to come which is retained because of a malpresentation

- the ewe is suffering from hypocalcaemia, a lack of calcium in the blood.

It is always a good idea to check for the presence of another lamb if a ewe has not passed the placenta after a few hours – ballotting the abdomen (see page 57) should help you to determine if a lamb is still to come. If so, carry out a careful internal examination, but **do not** pull the placenta. Doing this could cause the uterus to prolapse or could cause severe bleeding.

Providing there are no more lambs to be born, the ewe should be left to pass the placenta in her own time – this can take 2 or 3 days. To prevent infection and illness, appropriate antibiotic treatment, obtained from your vet, should be given. If the ewe is at all wobbly on her legs suggesting a lack of calcium, she can be injected with 50 ml of calcium borogluconate 20% under the skin over the ribs. Warm the calcium first by putting the bottle into a bucket of warm water. Use a 16 gauge needle and large syringe, inject quickly and rub in well, as it stings!

Fostering lambs

Most experienced shepherds in large flocks do quite a lot of swapping lambs around, in order to even up those ewes which have more lambs than they can rear with those which have fewer. The trick is in convincing the ewe that a lamb from another ewe is really hers! If you are not experienced, it is best to confine attempts initially to providing one lamb to a ewe which has lost her own at or very soon after lambing. Make sure the ewe is healthy and has plenty of milk before starting. Choose a reasonably strong foster lamb, as young as possible – a newborn one is fine, but dry it with a towel first. A weak lamb may not persevere with a suspicious ewe or may get trodden on and injured. Don't use a lamb of more than a week old, as the ewe won't have enough milk unless you give extra with a bottle until she is producing sufficient. It is very important to know that **if the fostering process fails, the real mother will not take the lamb back,** so you may end up with an orphan lamb!

Various methods involving confining the ewe in a yoke or tethering her with a rope or halter are used in large flocks. In general, whatever method is used, if the ewe will not accept the lamb after 2 to 3 days it is unlikely that the fostering will succeed. A successful fostering is of benefit to the ewe and lamb, providing that both are healthy.

The most useful methods for inexperienced shepherds or small flocks are as follows.

Wet fostering

This is particularly effective for a ewe whose lamb is born dead. Do not let the ewe stand up. Catch as much as possible of the birth fluids and membranes from the ewe in a clean container – a plastic washing-up bowl is ideal. Put the foster lamb in the bowl and wet it all over, if possible, with the fluids, particularly the head and tail region. Use any pieces of placenta to smear over the head and legs. Tie the lamb's legs together so it cannot stand up, or hold the lamb so it cannot stand up (this simulates a newborn lamb). Then put it gently in front of the ewe's nose with the hind end of the lamb presented to the ewe. Watch the ewe carefully – if she starts 'talking' to the lamb, all is likely to be well, so after a few minutes, both she and the lamb can be loosened and left quietly together. If the ewe is unwilling to lick the lamb, a handful of feed such as dry sugarbeet shreds scattered over the lamb may set her going.

Stimulation of the cervix

This method mimics the stretching of the cervix which takes place as a lamb is born and which is involved in bringing about maternal behaviour. The technique should only be carried out in a ewe which has lambed within the previous few hours, as it involves an internal examination, which **must** be carried out cleanly and gently. Put the ewe on her side and insert a well-lubricated hand, preferably gloved, into the vagina and back as far as the cervix. Make the hand into a fist, as this stimulates the ewe to start straining. Hold the fist in this position for a short time, then remove. Introduce the lamb, preferably smeared with any foetal fluids which were brought out on the hand, to the ewe. This method can also be used with success in cases where a ewe has refused or shown no interest in her own lamb.

Skinning the dead lamb

This is an age-old technique which often works well where a ewe has lost her lamb in the first few days. Cut the skin of the dead lamb around the legs above the knees and hocks, and around the neck. Next, cut the skin of the abdomen from the breastbone to udder or scrotum and separate the skin from the muscles, using the fingers to loosen it. Remove the skin, including the tail, by pulling, starting with the tail end (cut through the bones at the base of the tail as the skin is pulled) and finishing by pulling it over the neck. Rub the head and legs of the foster lamb with the skin, and then put the skin on it like a coat and adjust to fit if necessary, tying the 'jacket' with string under the lamb's belly (Figure 15.3). Carefully introduce the lamb to the ewe and watch for her reaction. Providing she does not attack the lamb, it can be left with her, but it is a sensible precaution to have a small creep area where the lamb can retreat if necessary. If this method is going to work, the ewe will accept the lamb within a few hours. The 'jacket' soon gets very smelly, but should not be taken off too quickly. Generally the tail part can be cut off within 2 days, removing the rest in one or two more pieces over another couple of days.

Figure 15.3
Fostering a lamb by using the skin of a dead lamb

CHAPTER 16

Postlambing prolapses

Uterus

Occasionally after lambing, which may have been assisted or natural, the uterus becomes prolapsed through the vulva. This may, at first, look like a prolapse of the vagina and cervix, but if it is examined carefully, it can be seen that the uterus has turned completely inside out (like a coat sleeve which has been turned inside out) (Figure 16.1). The diagnostic signs are:

* the ewe has lambed

* the presence of 'buttons' (cotyledons) where the placenta is attached to the lining of the uterus

* the likely presence of the placenta still attached at the cotyledons, although it is possible for a prolapse to occur after the placenta has been passed.

The longer the prolapse is out, the more danger there is to the ewe, as it may become damaged, swollen and chilled. It should, therefore, always be treated as an emergency. Even though this condition looks serious and impossible to correct, results are usually good providing treatment is given early.

* Restrain the ewe on her side in a safe place (not in a small pen in case the prolapse is damaged on a hurdle).

* Wrap the prolapse carefully in a clean damp cloth to protect it.

* Keep the prolapse damp by pouring warm water gently over it until it can be replaced.

Replacement is a job for the vet, particularly if the prolapse has been out for some time.

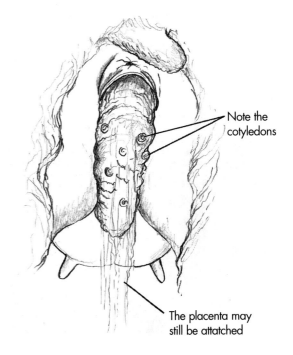

Note the cotyledons

The placenta may still be attatched

Figure 16.1
Prolapsed uteris

A pain-relieving injection can be given (only by the vet) into the spinal canal, which, as well as making replacement easier, will relieve straining for 24 hours or more, allowing contraction of the uterus to take place.

You should not attempt to replace the prolapse yourself unless you have had instruction from your vet, or in exceptional circumstances where it is not possible to get expert help. If it is necessary to deal with it yourself, gather together the following equipment:

* hoist

* 3 pieces of towelling (one of which is large enough to wrap around the uterus)

- a piece of cloth not less than 30 cm square

- 2 clothes pegs

- 2 buckets of warm water (body temperature)

- soap solution

- antiseptic cream (several intramammary tubes will do, but do not use them if you are allergic to penicillin)

- prolapse harness

or

- suture material, eg. tape 6 mm wide and about 60 cm long

- suture needle with large eye (preferably a Buhner needle which has a handle)

- some short lengths of polythene or gas tubing

- clean scissors

1. Put the ewe on the hoist and raise her about about halfway.

2. Remove any loose straw and the like from the uterus, then wash gently with soap solution and rinse using a piece of towelling.

3. If the placenta is loose, remove it; otherwise leave in place, as attempting to remove it can cause serious bleeding.

4. Discard the dirty water, then use another bucket of clean warm water and a clean piece of towelling to give a second rinse and to warm the prolapse.

5. Wrap the uterus using the large piece of dry towelling. Slide the piece of cloth like an apron under the uterus, isolating it from the anus and any dirty wool.

6. Hold the piece of cloth in place with a clothes peg either side of the ewe's hips – this keeps the area clean during replacement.

7. Raise the ewe fully until only her shoulders are resting on the ground.

8. Gently remove the towelling from around the uterus.

9. Coat the uterus with antiseptic cream.

10. An assistant stands astride the ewe, placing the hands (clean!) either side of the vulva to form a funnel.

11. Lift the uterus into the 'funnel', then, using the flat of the hand only, gradually work the uterus back into place in between the ewe's straining.

12. After the uterus goes back in the abdomen, follow in with the hand, then, making the hand into a fit, push gently back into place, checking each horn to make sure they are not still inverted (if they are not back in place the ewe is likely to attempt to prolapse again).

The prolapse must next be retained in place.

- If the prolapse was small and the ewe is not straining too much, she can be carefully lowered and a well-fitting prolapse harness put immediately in place. This can be removed after 7 to 10 days.

- If the ewe is straining a lot, there may, in the absence of the vet, be no alternative to stitching the vulva, but this should only be done if you are confident of being able to carry it out without causing the ewe unnecessary pain. Leave the ewe on the hoist whilst doing the stitching. See pages 27-29 for details on the stitching technique.

Aftercare

- Antibiotics are necessary for several days to prevent infection – consult your vet.

- The ewe may be short of calcium, so inject 50 – 100 ml calcium borogluconate 20% under the skin over the ribs after replacing the prolapse. Warm the calcium first by putting the bottle into a bucket of warm water. Using a large syringe with a 16 gauge needle, inject quickly and rub in well after.

- Give a single dose of 100 ml liquid parafin by mouth to avoid straining to defaecate.

- Mark the ewe so she can easily be identified.

- Make sure the ewe gets plenty of exercise. She can be returned to the rest of the flock unless she is unwell.

- If the placenta is still in place when the prolapse is replaced, this will loosen and come away after a day or two. If it gets caught up in the harness, loosen this to allow the placenta to come out. Do not on any account pull on the placenta or this may cause the uterus to prolapse again.

- The harness or stitches can be removed after about 10 days.

Cervix

Although prolapse of the cervix is most common before lambing, it can also occasionally happen after lambing, sometimes days or even weeks later. The cause is usually damage to the cervix at lambing, perhaps tearing whilst attempting to dilate a case of ringwomb or in lambing a very large lamb. In spite of antibiotic treatment the cervix may become infected and thickened, stimulating the ewe to begin straining. The cervix appears at the vulva where, if it is not noticed for a day or two, it becomes dried and covered with faeces. Unlike a prelambing prolapse, the urethra is not usually obstructed, so passing urine is still possible. The ewe is therefore less troubled and may behave normally between the odd bouts of straining. The prolapse must be dealt with, for if it is not, it will become increasingly infected and, in warm weather, will become a target for fly strike. The longer the time since lambing has elapsed, the more difficult it will be to return the prolapse to its correct position, since the pelvic ligaments will have tightened and it may be difficult or impossible to insert the hand into the vagina.

Raise the ewe's hindquarters. Carefully clean the prolapse with warm water and soap solution. This may take some time because of the dried-on faeces. When it is clean, lubricate the prolapse well with antiseptic cream and, with the palms of the hands, push the prolapse back until the cervix is in its normal position. If it is impossible to insert the hand into the vagina, the case of a large syringe can be used to assist in this. The prolapse must then be retained in position.

- Providing the ewe is not straining badly, a well-fitting prolapse harness can be used. Check daily to make sure no parts are rubbing and injuring the ewe.

- If the ewe is straining, it may be necessary to insert a stitch. Preferably this should be done by your vet, who will give an injection into the spinal canal to control straining.

Whichever method is used to retain the prolapse, it should be left in position, preferably until the ewe is culled. If the harness or stitch is removed, even after several weeks, it is possible for the prolapse to recur; therefore the ewe should be clearly marked and culled as soon as possible. She should definitely not be retained for further breeding.

CHAPTER 17

Essential Notes on the Newborn Lamb

After a normal lambing, the ewe usually immediately gets to her feet and turns to start licking the lamb which shakes and raises its head. These actions remove membranes and excess fluid from the face of the lamb and stimulate the lamb to start breathing normally. If the ewe does not get up and the lamb's face is covered with membranes, the lamb will quickly suffocate, so the face should be cleared of membranes and the lamb resuscitated immediately. If the ewe and lamb are behaving normally, it is better to leave them undisturbed at this stage, allowing the ewe to finish lambing in peace. However, if there is a danger of other ewes interfering or even trying to steal the lamb (newborn lambs are a source of great interest to other expectant ewes!), if several ewes are lambing at the same time, or if there is a danger of the lamb wandering away, it is better to move the ewe and lamb to a close-by individual lambing pen, or to put up a temporary pen where the ewe should be observed to make sure that lambing is completed.

Resuscitation

This should be carried out for any lamb which is seen to be in distress (this particularly includes excessive kicking or rolling movements) or not breathing properly. Lambs born backwards, including breech births, are often distressed at birth and may be difficult to revive.

Establishing breathing

- Clear the nose and mouth of membranes and fluid. A clean piece of towel is ideal for this.
- Check for a heartbeat by feeling the chest between finger and thumb. If there is no heartbeat, the lamb is dead and further efforts will be fruitless.

- Holding the lamb by its back legs with the head down, clear any further fluids from the nose and mouth.

- If the lamb was born backwards, there may be excess fluid in the nose and throat; therefore carefully swing the lamb in a complete arc horizontally or a half arc vertically which helps to clear this away. Hold tight as the lamb will be slippery! Any swinging must not be done violently or the lamb may be damaged. Again, clear away any fluids from the nose.

- Briefly lay the lamb on its side and **lightly** slap the chest. Observe for a few seconds to see whether breathing is starting.

- If not, administer some respiratory stimulant drops (from your vet) under the tongue as directed on the bottle.

Other methods are listed below.

- Lay the lamb on its side. Get hold of the front legs with one hand and the back legs with the other and push slowly towards, then away from each other. Repeat several times.

- Inserting the end of a piece of straw just into the entrance of one of the nostrils is an old-fashioned method of stimulating sneezing and breathing which often works.

- Twisting the lamb's ear may stimulate breathing.

- Use an acupuncture point, which is in the line between the nose and mouth – stimulate by pressing hard on this line with a fingernail.

- If breathing still does not become established, feel the ribs to see whether they are broken and do not carry on this procedure if they are. Lay the lamb on its side, make sure the airway is clear, and insert the first and second fingers into the side of the lamb's mouth and open the fingers to keep the jaws apart. With the palm of the other hand, **gently** press on the lamb's ribs for one second to compress the lungs, then with a quick, sharp movement, take the hand away for two seconds to allow the lungs to take in some air. Repeat until the lamb is breathing unaided. In obstinate cases turn the lamb over and repeat on the other side.

- If an assistant is available, the lamb can often be stimulated at the same time by **lightly** flicking it all over with the fingertips, stimulating the quick licking movements of the ewe's tongue.

- Various commercial devices are available which may assist in starting breathing in difficult cases. 'Mouth to mouth' resuscitation should **not** be used for lambs as there is a risk of spread of some diseases, for example some types of abortion, from lambs to people.

There is a method of inflating the lungs of the lamb with a stomach tube, which may be used if the lamb cannot expand its chest (it goes in rather than out when it tries to take a breath) and looks likely to die. It is possible to practice the technique on a dead lamb, so that it can be quickly applied when needed in a real emergency.

- Place the lamb on its right side. Pass a stomach tube into the mouth and into the first part of the oesophagus – it will be possible to feel the tip of the tube if the underside of the neck is pinched between the finger and thumb of one hand. Keeping the finger and thumb in position, which closes off the oesophagus, slightly withdraw the tube so that it lies in the back of the mouth. Then cup the other hand over the mouth and nose, to make an airtight seal around the tube. Inflate the lungs by blowing through the tube until you see the chest rise. Then remove the tube. This method will not save all lambs which will not breathe, but is worth a try.

If the lamb dies within a few minutes in spite of all attempts to save it, there are several possible causes:

- lack of oxygen during birth, leading to irreversible brain damage

- damage to the brain during birth eg. haemorrhage

- immaturity of the lungs, which are unable to inflate due to lack of surfactant (a natural substance produced in the lungs of the foetus in the late stage of development. In lambs this is not produced until very near to full term)

- fracture of the ribs during birth, which has resulted in severe lung damage

- rupture of the liver, leading to fatal haemorrhage – this may be caused by pulling too hard on the lamb as it is being born or may be spontaneous as a result of too little vitamin E in the ewe's diet

- a serious congenital defect such as abnormal development of the heart or brain.

Routine checks for a newborn lamb

Once the immediate concern of getting the lamb to breathe has been dealt with, there are several other checks which should be carried out to ensure that all is well with the lamb.

Is the lamb breathing regularly and evenly?

If it is not, it could be the result of brain damage and oxygen deprivation, or because of cracked or broken ribs.

Look at the chest and gently feel the ribs – affected lambs have a 'caved-in' appearance of the chest on the affected side. Providing the damage is not too bad, most will survive if handled very gently to avoid further damage. The ribs heal very quickly.

Does the lamb shake its head and sit up quickly?

This is a sign that the lamb is probably normal and healthy. If it does not, the lamb may have brain damage as a result of a lack of oxygen or even haemorrhage over the brain caused during a difficult delivery; this is particularly likely if the lamb is large. If the lamb is small, it may be immature or premature especially if the ewe is undernourished or thin. Lambs affected by diseases such as swayback, border disease or daft lamb disease (see page 82) also fail to progress normally in the period immediately after birth.

Any lamb which does not quickly sit up and try to stand requires extra attention. Even if breathing is established, they may lie around, rapidly becoming cold and will fail to suck. These lambs require extra care – feeding with colostrum (see page 67) and a source of warmth (see page 71).

If a lamb does not get up quickly, it may need protection from the ewe who may accidentally tread or lie on it (see page 72).

Is the navel cord normal?

It is good practice to treat the navel cord with iodine or an antibiotic spray, particularly for lambs born indoors; this speeds up drying and helps to guard against infection. Take care that the whole length of the cord is treated, particularly the end, where infection can most easily enter.

Occasionally the navel cord will bleed abnormally after birth. This should be stopped as the lamb may otherwise bleed to death. Try the following methods of stopping the bleeding:

- With **clean** fingers, apply pressure to the cord just below where it joins the skin, by pinching between the finger and thumb. At the same time, run the fingers of the other hand down the cord to squeeze out any blood within it. After a minute or so, release the cord and watch carefully to see if it fills with blood again. If not, spray the cord well with antibiotic spray. Check again after a short time. The cord should be sprayed again the following day.

- If the bleeding does not stop with finger pressure, tie a sterilised thread or put a sterile clamp just below where the cord joins the skin, squeeze out any blood from the cord and spray well with antibiotic spray.

Occasionally the intestines may prolapse through the navel (see page 79).

Eye check

If entropion (inturned eyelids) is a problem in the flock, eyes should be routinely checked and corrected if necessary (see page 80).

Moving lambs

Lambs should not be carried by their front legs only. If they are to be carried, a hand should be placed between the front legs under the breastbone to support the weight of the body. Ewes which are experienced mothers will usually closely follow their lambs from the field or pen. Very flighty or inexperienced ewes may be less willing to follow and need to have constant contact with the lamb if they are not to be frightened away. A simple disposable sling-type of carrier can be made from a paper feed sack (Figure 17.1), which enables the lamb to be carried in a natural position near to the ground so that the ewe can follow with her nose near the lamb. This carrier also helps to preserve body heat if the lamb has to be moved some distance.

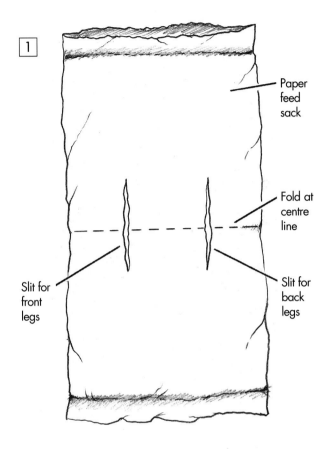

Figure 17.1
Making a carrier for a newborn lamb

1

Paper feed sack

Fold at centre line

Slit for front legs

Slit for back legs

The above carrier should be used only once to avoid cross infection.

2

Carrier ready for use

3 Two ends of paper feed sack rolled together to secure the lamb, and give a firm grip

Colostrum and feeding

It is essential that the lamb gets colostrum as soon as possible after birth. This provides:

* energy to keep the lamb warm and to begin the growing process (a healthy lamb grows very rapidly, 250–400 g/day is not unusual)

* antibodies to protect against disease – these are produced by the ewe in late pregnancy in response to environmental micro-organisms

and to any vaccines which have been given (clostridial vaccination is essential 4 to 6 weeks before lambing, other vaccinations depend on individual flock problems).

If a lamb is strong, attempts to stand immediately, starts looking for the teat and the ewe has plenty of colostrum, all should be well. The ewe and lamb(s) can be left to get on with things naturally, although they should always be checked after an hour or two to make sure the lambs look full and contented (see also 'Watery mouth' page 79, as this will also make them look full, but miserable rather than contented).

At-risk lambs

If in doubt, give some colostrum! It is often stated that a lamb should get colostrum by 6 hours after birth but **this is far too late** – by this time a lamb may already have succumbed to hypothermia if the weather is cold and wet or windy. Those most at risk are:

- small

- weak

- very large and dozy

- general lack of 'get up and go'

- from a litter of more than two lambs

- from a ewe which does not appear to have sufficient colostrum

- from young inexperienced ewes

- from thin ewes

- from old ewes

- from sick ewes

- from ewes with very thick colostrum or large teats.

What to give

- Colostrum from the lamb's mother is best if she has enough. Milk any surplus into a plastic jug and transfer to another jug at intervals so that if the ewe kicks you don't lose all of it.

- Colostrum from another recently lambed ewe is also suitable. Make sure her own lamb(s) have had their fill first.

Surplus ewe colostrum can be frozen and will keep for at least a year. It is worth milking any ewe with excess colostrum on the first or second day. Try to keep enough in store so you have some left over for the beginning of the next lambing season. Freeze in small containers such as yogurt pots which hold 100 ml, enough for one lamb feed, or ice cube bags which hold about 15 ml each and can be quickly defrosted. Thaw by putting the container in hot (not boiling) water, or gently heat in a double pan, not on direct heat which will curdle and spoil the colostrum. Stir frequently to break up any large lumps and to ensure even heating throughout to body temperature. Colostrum can be thawed in a microwave oven, but great care has to be taken that it doesn't heat unevenly or overheat as this will spoil it.

Substitutes for ewe colostrum

1. Goat colostrum is very similar to ewe colostrum and is an excellent substitute. Maedi-accredited flocks should use only colostrum from goats free of a closely related disease, caprine arthritis-encephalitis (CAE). For lambs going for meat, colostrum from untested goats will do, since the lambs will be gone long before there is any risk of disease developing.

2. Cow colostrum can also be a good substitute, although it doesn't contain as many antibodies against sheep diseases. The main problem is that the occasional cow produces antibodies in its colostrum which can damage lambs' red blood cells leading to anaemia at 1 to 3 weeks old. Try to avoid this by saving colostrum from several cows and mixing it together before feeding. Freeze from individual cows until you have enough, then allow all to thaw gently at room temperature, mix, divide into suitable containers and re-freeze. (Although repeated freezing and thawing is not a good idea, it is the most practical way of dealing with this particular problem and, if you are careful, will not substantially damage the antibodies.)

3. Commercial dried colostrum powders are useful to have as a backup to add to a limited amount of ewe

colostrum. If you have to rely entirely on this type of product, each lamb will need several packets.

Dried milk powders are **not** suitable for feeding lambs in their first day. After the first day, specially formulated lamb milk powders are suitable both for rearing or topping up lambs. Cows' milk is not suitable as it is much lower in fat than ewes' milk; lambs reared on cows' milk generally do not thrive very well and often have a 'pot-bellied' appearance.

How to give the colostrum

Holding a weak lamb on the ewe's teat trying to get it to suck is guaranteed to try the patience of even the most dedicated shepherd! It is also useless as regards knowing how much colostrum the lamb has sucked. The lamb can be fed with a bottle and teat but this can take a lot of time and patience and both may be in short supply. The most effective way is to feed with a stomach tube – this is a quick and safe method, particularly for the first feed, and will usually ensure that most lambs get on their feet and are off to a really good start.

• Be cautious about using a stomach tube if the lamb has been with the ewe for long enough to suck by itself as the stomach could already be full. In this case it is safer to offer a bottle and teat so the lamb can take as much or as little as it needs.

• Also take care if the lamb is very weak and cannot properly hold its head up. Preferably treat as for hypothermia (see page 77). Otherwise, give very small amounts, say 20 ml at frequent intervals until the lamb is strong enough to sit up.

Using stomach tubes

These are made of plastic or rubber. The plastic type is rather stiff, particularly in cold weather, so the end should be dipped in warm water or milk before use. Colostrum is given via a funnel or large

syringe (which can be used as a funnel by removing the plunger) attached to the tube (Figure 17.2). A funnel has the advantage that colostrum runs into the lamb's stomach at its own speed, so it should not be possible to injure the lamb. The disadvantage of a funnel is that some colostrum is so thick that it will never run through! A syringe saves time, and providing it is used carefully and slowly, should be safe for the lamb. Tubes and funnels or syringes should be washed after each use in soapy water and rinsed well afterwards. Disinfectants can be used, particularly if lambs are sick, but these sometimes damage the tube and particularly the syringes.

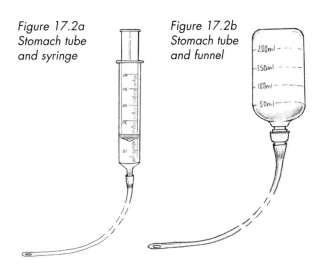

Figure 17.2a
Stomach tube and syringe

Figure 17.2b
Stomach tube and funnel

200ml
150ml
100ml
50ml

If you are unsure about how far to put the tube, measure it against the lamb from the mouth to behind the last rib before using, then use a marker pen to indicate the correct length.

• Lay the lamb across your knee or, if it is wet, on a straw bale.

• With the left hand hold the lamb's head by grasping it from above in front of the ears.

- Lift the lamb's head so that the neck is stretched out.

- Hold the tube in the right hand and dip the end of the tube in warm water or milk to lubricate and soften it.

- Insert the tube into the centre of the lamb's mouth (Figure 17.3a) and gently push over the tongue. Carry on gently pushing the tube – the lamb will swallow and it should go into the oesophagus without resistance. If the lamb coughs and splutters or shows discomfort, withdraw and try again.

Figure 17.3a
Passing a stomach tube

- Now pinch the underside of the neck between your finger and thumb – you should be able to feel the end of the tube through the skin (try moving it gently backwards and forwards if you cannot feel it). You should also be able to see it moving down the left side of the neck unless the lamb has very thick wool.

- Once you have felt the tube in position, continue pushing until the required length of tube is down (if it is not pushed far enough there is a danger of the colostrum being regurgitated and getting into the lamb's lungs). When the tube enters the lamb's stomach you may hear a characteristic gurgling sound.

- Providing the lamb is not distressed, attach the funnel or syringe (Figure 17.3b). Either allow the warm colostrum to run in by gravity, or gently and slowly push the plunger to empty the syringe. Refill the funnel or syringe until the lamb has had the required amount.

- Pinch the tube tightly during removal so that any colostrum remaining in it does not get inhaled.

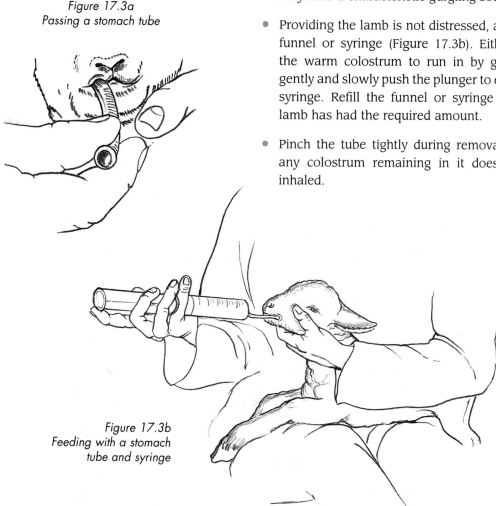

Figure 17.3b
Feeding with a stomach tube and syringe

How much colostrum to give

● Give a first feed of 100 ml to an average-sized lamb, 150 ml to a large lamb and 50 ml to a very small lamb. Make sure the lamb has had at least 50 ml/kg bodyweight of top-quality colostrum by the time it is 2 to 3 hours old.

● If the lamb is not with the ewe, but is intended for fostering, it can be given more feeds of colostrum by stomach tube until a ewe is available. It will need about 200 ml/kg bodyweight during the first 18–20 hours of life split into 4 or 5 feeds.

● If the lamb is left with the ewe and has the opportunity to suck her, it is better to offer a bottle after the first feed so the lamb can take as little or as much as it needs.

Feeding after the first day

Healthy lambs with healthy ewes with plenty of milk should require no more attention, other than regular checking to make sure all is well.

● Orphan lambs should go on a bottle and teat or automatic feeder, even though this can make fostering more difficult, as continued use of the tube can make the throat sore. Milk substitute can be fed according to the maker's instructions.

● Lambs remaining with a ewe which has limited amounts of milk should be offered milk substitute from a bottle 2 or 3 times a day.

● It is good practice to offer a bottle to any lamb which appears hollow or hungry in the first few days to tide it over until the ewe is producing enough milk. This may forestall the lamb becoming a poor doer.

● Older lambs which are short of milk, eg. if the dam is ill or dies, can be extremely difficult to deal with as they often refuse to suck from a bottle. Great patience and perseverance are

needed but most will eventually get the idea. You should offer water and palatable creep feed to encourage them to eat as soon as possible. Lambs cannot manage without milk until between 4 and 5 weeks of age at the earliest.

Shelter and warmth

Preventing lambs from becoming hypothermic (which happens if they lose a lot of heat because of bad weather, or starve because they are not getting enough milk) is an important part of maximising lamb survival (for methods of dealing with hypothermic lambs see page 77).

Lambing in a field with natural shelter will be adequate in all but the worst weather for ewes which are good mothers, have plenty of milk and have singles and strong twins. If natural shelter is poor, other means of shelter such as straw bales should be provided.

Lambs born indoors are less likely to become chilled, but may do so if they are short of milk or the building is draughty and cold.

● If the ewe does not lick the lamb, dry it with a clean towel. This is particularly important if the weather is cold and windy, as a weak lamb will soon become chilled. (Give some colostrum by stomach tube as well.) An effective temporary shelter can be made from a paper feed sack turned inside out (to avoid the lamb inhaling dusty remnants of feed). Place the lamb in the sack with the head at the open end and with the closed end towards the wind and weather. A fresh sack should be used for each lamb.

● You should have some pens prepared where extra heat can be provided for weak and small lambs. It is a complete waste of time (and dangerous) to suspend a heat lamp above a pen at ewe height – she will stand under the lamp and get her fleece burnt and the heat does not get to the lamb which needs it. A simple creep area can be constructed (Figure 17.4) where the lamb can benefit from the heat and the ewe cannot get to it. You will need a piece of strong wire mesh approximately 1m by 60 cm and some baler twine.

1. With the mesh held vertically, put it across an individual pen so that the long side forms a triangle across one corner of the pen.

2. Lift the piece of mesh so that the bottom edge is about 50 cm from the ground (make sure there are no sharp projections on the lower end of the mesh).

3. Tie the mesh firmly in position to the two sides of the pen using baler twine.

Figure 17.4
Lamb creep area

4. Suspend a heat lamp in the triangular creep area thus formed. The lamp should be 60–70 cm above the ground and should be suspended by a chain from a beam above so that it cannot accidentally fall and cause a fire.

The creep area provides heat for the lamb(s) and also helps to prevent overlying by a clumsy ewe since the lambs quickly learn to go and sleep there in safety.

- The creep area can be modified for a lamb which is in danger of being trampled by the ewe by putting a second piece of mesh across at ground level to confine the lamb, but leaving it within sight and smell of the ewe. If this is done, the lamp **must** be raised considerably higher (to about 1 m – check how warm it feels down at lamb level) to prevent the lamb becoming overheated, since it cannot escape from the creep area. The lamb must then be put to suck the ewe regularly and can be released from the creep as soon as it is strong enough.

Castration and tailing

Although these procedures are often carried out as a matter of course, think carefully whether they are absolutely necessary as they inevitably cause some pain to the lamb. Castration is not necessary if the lambs are to be sold for meat before about 6 months of age, as uncastrated males grow faster and produce leaner carcases. If ram lambs are likely to be around for longer than 6 months it is safer to castrate them so as to avoid unwanted pregnancies which can cause many problems and result in losses of both ewes and lambs. Tailing may be necessary if the lambs are prone to scouring (some breeds are worse than others for this, even if worms are well controlled).

If it is necessary to tail and/or castrate, the easiest method is by tight rubber rings applied with an applicator called an elastrator. The rings should not be applied before 24 hours to allow the lambs a good start, but must be done by UK law before 7 days, and can only be applied by people of 18 or over. Try to get someone experienced to show you how to do these jobs, as it is possible to put the rings in the wrong place if you don't know what you are doing!

Castration

Have an assistant holding the lamb with its belly towards you. Feel the scrotum to make sure that both testicles are present. If you cannot find both, or if a hernia is present (the scrotum is larger than normal and contains part of the intestine) do not proceed any further. Mark the lamb (eg. leave with a long tail if the rest are tailed) and try to make sure it is fattened before about 6 months old.

Figure 17.5a
Castrating a lamb with a rubber ring

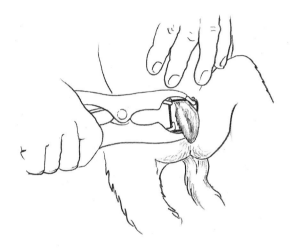

Figure 17.5b
Castration ring correctly placed

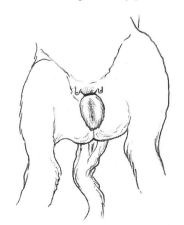

- If both testicles are present, hold the applicator with the prongs towards the lamb. Use one hand to apply slight pressure above the testicles to ensure they are not withdrawn into the abdomen. Open the jaws of the applicator and place the open ring over the scrotum. Allow the ring to close above the testicles, and leave on the applicator until you are sure the ring is in the correct position above the testicles but below the teats. Check that the lamb's penis is not trapped in the ring – the lamb will be unable to pass urine and will die if this happens. When you are sure the ring is correctly placed, work it carefully off the prongs and check the positioning again (Figure 17.5).

Tailing

If the lamb has lumps of faeces around the tail, remove these carefully first.

- Place the ring on the tail with the prongs of the applicator facing the lamb. Put the ring on so that enough tail will be left to cover the anus of a male or the vulva of a female (Figure 17.6). In practice, this means putting the ring just below the triangular hairless area under the tail for a male and at least 1.5–2 cm lower for a female.

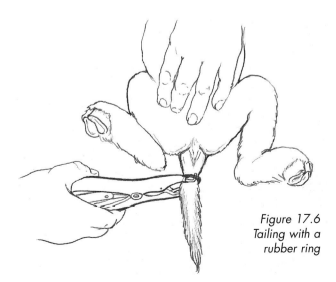

Figure 17.6
Tailing with a rubber ring

After castration and/or tailing, the lamb will show signs of pain, rolling and lying stretched out, often for about an hour. These procedures are painful, but there is no good way of relieving this at present. It is possible that less painful methods may be developed. The message is, think carefully whether you need to do these procedures in the first place.

CHAPTER 18

Keeping the Lambs Alive

An observant shepherd will automatically check the health of young lambs as part of the daily routine. However, even in the best-managed flock, problems and losses do occur. Surveys in the UK have shown that approximately a third of all lamb deaths are due to abortions and stillbirths, another third to hypothermia, and the rest are made up of a variety of factors including defects present at birth, infections, accidents and predators.

Here are some simple checks which can be made to try to determine the possible cause if a lamb appears to be unwell. More information about some of the more common, important, or emergency conditions mentioned is given in the next chapter.

Lamb examination check-list

Does the lamb stretch when it stands up?

- If it gives a good stretch and appears bright and alert there is unlikely to be much wrong.

What is the lamb's temperature?

- Normal is 39–40^{0}C.
- Higher than 40^{0}C indicates infection.
- Lower than 39^{0}C indicates hypothermia (see page 77).

What is the appearance of the lamb's anal region?

- If the tail is stuck down by sticky orange faeces this indicates the lamb has had lots of colostrum (a good thing!). Gently remove the lumps of faeces and make sure the tail is free.
- If the lamb is scouring, see page 81.

- If the lamb has no anus (atresia ani), see page 80.

Can the lamb stand properly?

- If the lamb shakes itself frequently, this indicates that it is premature or immature and will need extra care for the first few days.
- If the lamb is shaking continuously and has an abnormally hairy coat (hairy shaker), it is probably affected with border disease.
- If the lamb cannot stand, or has difficulty standing and throws its head abnormally back, it may be affected with daft lamb disease.
- If the lamb is floppy and unable to stand, it may be suffering from vitamin E deficiency or severe swayback.

Is the lamb off its back legs?

- If it cannot stand properly, the possible causes include swayback, spinal abscess or joint ill (see page 81-2).

Are the lamb's legs and joints normal?

- If the lamb cannot straighten its front legs properly, it is suffering from contracted tendons (see page 81).
- If some of the joints, especially knees and stifles, are swollen or painful, the lamb has joint ill (see page 81). Affected lambs usually look tucked-up and sorry for themselves as well as being stiff or lame.
- If lame on one leg, it may have a fracture – young lamb's leg bones are easily broken, but also usually mend easily, particularly if the break is below the knee or stifle. If the leg is not

out of line, a well-padded splint may be all that is necessary. If the break is not straightforward, see your vet for treatment.

- If lame, another possibility is a swollen claw, caused by an abscess. This needs bathing to remove pus, as well as antibiotic treatment.

Is the breathing normal?

- Rib damage can be caused by a clumsy ewe, as well as during birth.

- If the breathing is very fast and shallow and the lamb is weak, this may be another indicator of vitamin E deficiency (white muscle disease).

- If the breathing is fast and the lamb's temperature is above normal, the lamb may have pneumonia.

Is the mouth normal?

- Check for jaw damage, or missed congenital defects such as cleft palate.

- Excess salivation indicates watery mouth (see page 79).

- If the lips or gums are swollen and bleeding, the lamb probably has orf. Handle with great care wearing disposable gloves, as this can cause disease in people.

Are the lamb's eyes normal?

- If the eyes are watering excessively or the surface of the eye looks inflamed, the lamb probably has inturned eyelids (entropion, see page 80).

- If the eyes are sunken, the lamb is dehydrated and will need feeding with electrolyte solution.

What is the shape of the abdomen?

- Is the abdomen full or tucked up and empty?

- If full, is it milk or gas? Tap gently – gas sounds like a drum. This probably indicates watery mouth (see page 79).

- If empty, the lamb may simply be short of milk, or may have other problems as well. Try offering a bottle if no other abnormality is found.

What is the appearance of the navel?

- If the navel is still wet in a lamb older than one day, reapply navel dressing.

- If thickened or painful, it is infected (navel ill) and needs antibiotic treatment.

If you are unable to make a diagnosis, lambs do not respond to treatment or many are affected, you should consult your vet.

CHAPTER 19

Problems and Emergencies in the First Few Days

This chapter brings together some information about the most common emergencies and problems which may occur during the first few days. Detailed information about these diseases and problems, and others occurring after the first few days, will need to be sought elsewhere.

Hypothermia

Chilling and starvation kill many lambs each year unnecessarily, although sudden bad weather can catch even the best shepherd (or weather forecaster!) unaware. Making sure that ewes have plenty of colostrum and milk for their lambs, that the lambs are well mothered up and that there is plenty of shelter are all important parts of preventing hypothermia.

Most hypothermic lambs will have used up much or all of their energy reserves and will have a very low blood glucose concentration, so it is important that they are given some source of energy before warming. If this is not done, the lamb is likely to go into fits and die.

Treatment of hypothermia depends on how badly affected the lamb is. If it is only mildly affected, it may be possible to leave the lamb with the ewe, but if it is very cold and weak, it will have to be removed for treatment. If the ewe has another lamb, and you are intending to return the chilled lamb to the ewe, take both away and return both together, as otherwise she will probably not take the weak one back. Alternatively, accept that if you revive the chilled lamb it will probably have to be fostered or bottle reared.

How badly affected is the lamb?

- Take the temperature of any lamb which looks miserable or is unwilling or unable to stand. You cannot accurately tell how cold a lamb is by feeling its mouth or ears.

- Normal temperature range is 39–40^{0}C.

- Once a lamb's temperature starts to fall, action must be taken or the lamb will progressively worsen and eventually will die.

- If a lamb's temperature is 37^{0}C or less, it is severely hypothermic and needs urgent treatment. An easy way to remember this is that if the lamb's temperature is at or less than normal human body temperature (37^{0}C), the lamb is in severe danger and needs immediate treatment.

What to do if the lamb can hold its head up

DRY ➤ FEED ➤ WARM ➤ FEED ➤ OBSERVE

Do not try to heat the lamb without first supplying energy – you may kill it!

- If the lamb is wet, dry it with a clean towel.

- Give 50 ml/kg bodyweight of good-quality warmed colostrum by stomach tube.

- If the lamb is mildly chilled, confining to a heated creep area as described on page 72 may suffice. Supervise to make sure that the ewe has sufficient milk and that the lamb starts sucking properly.

- If the lamb is moderately or severely hypothermic, remove it from the ewe and provide warmth. A warming box is safest, but other methods can be used. One useful method, **although it needs close supervision**, is to use a metal dustbin with a heat lamp suspended over it at a **safe** height. Turn the lamb frequently so it warms evenly.

- **Whatever method is used a close watch must be kept to see that the lamb does not get overheated. Check its temperature every 10–15 minutes and remove from the heater as soon as it is near normal.**

- Feed again. Then if it is strong enough, return to the ewe and keep under close observation.

What to do if the lamb cannot hold its head up

INJECT GLUCOSE ⟶

DRY ⟶ WARM ⟶ FEED ⟶ OBSERVE

Do not try to heat the lamb without first supplying energy – you may kill it! Do not feed by stomach tube until it can sit up as the milk may be regurgitated and inhaled into the lungs.

You will need:

- some 20% glucose (dextrose) solution. You may be able to get commercially prepared 40% strength, which will need diluting with an equal volume of freshly boiled water before use. Alternatively, in an emergency, you can make your own by adding 20 g glucose powder to 100 ml freshly boiled water. Remember, this is going to be injected into the lamb, so you must be scrupulously clean when preparing and handling

- a sterile 50 ml syringe

- some sterile 20 gauge x 2.5 cm needles.

1. Remove the lamb from the ewe.

2. Fill the syringe with glucose solution. **This must be 20% strength, sterile, at body temperature.** The dose is 10 ml/kg bodyweight.

3. Hold the lamb by its front legs, with the back legs hanging down.

4. Spray an area 2.5 cm behind and 1.5 cm to one side of the navel with antibiotic or antiseptic spray.

5. Attach a sterile needle to the syringe and insert the needle into the abdomen through the disinfected skin. The needle should go fully in at about 45^0 pointing backwards towards the lamb's hip (Figure 19.1).

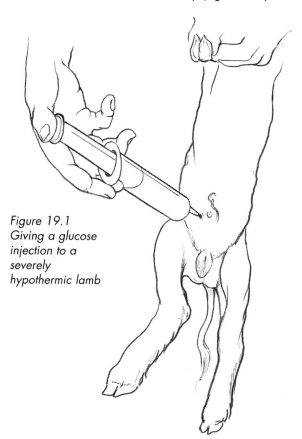

Figure 19.1
Giving a glucose injection to a severely hypothermic lamb

6. Slowly inject the glucose into the abdomen. The lamb may kick and bleat during this procedure. It may also pass some urine – no, it isn't the glucose coming straight out!

7. If the lamb is wet, dry it with a clean towel.

8. Warm the lamb, checking its temperature frequently.

9. When it can lift its head, feed with some good-quality warm colostrum by means of a stomach tube.

10. When its temperature is normal and it is strong enough, feed again and inject with antibiotic.

11. Return to the ewe if possible and keep under close observation, as the lamb may become hypothermic again if it fails to suck properly.

It is unlikely that you will succeed in reviving all hypothermic lambs, but it is well worth having a go. Better still, try to prevent them becoming hypothermic in the first place by making sure they have a full stomach and adequate shelter.

Watery mouth (rattle belly)

This is a disease of intensification, which has become much more common in recent years. Affected lambs, usually 1 to 2 days old, become dull and depressed and drool saliva. The abdomen is, at first, empty and tucked up, but then fills with gas (this can make the casual observer think the lamb is full of milk). The lamb progressively becomes weaker and colder, and rapidly becomes comatose and dies.

It results from lambs swallowing bacteria (in the environment, on ewe's wool or teats) as they try to find the teat to get their first feed of colostrum. The bacteria pass unhindered to the gut, where they multiply and produce toxins which eventually will kill the lamb. The most important thing to realise is that lambs are much less likely to get watery mouth if they have had a good feed of colostrum soon after birth (minutes, not hours!). Antibodies in the colostrum kill off the bacteria before they can do any harm.

Treatment of affected lambs is time consuming and not very successful, unless they are spotted at a very early stage. Antibiotics by mouth and injection are needed, as well as feeding with electrolyte solution and good nursing. See your vet about treatment, but better still, try to prevent lambs becoming affected in the first place.

- Make sure ewes are in good condition so they produce good strong lambs and plenty of colostrum.

- Make sure all lambs get colostrum within 15 minutes of birth – use the stomach tube to feed any suspect lambs.

- Keep lambing areas as clean as possible, whether indoors or outdoors.

- Do not tail or castrate lambs too early – this can put them off sucking at a crucial time. Leave until the lambs are well on their feet if you have to do these procedures.

- If, in spite of having applied all these preventive measures, you still have a problem, your vet may recommend dosing lambs at birth with antibiotic, but this is not a desirable practice in the long term.

Herniation of intestines

This is a sporadic emergency which happens soon after birth. For some reason – perhaps the ewe is overzealous at licking the lamb, or perhaps because of a weakness in the abdominal wall – the intestines of the lamb come out through the navel. At first the navel cord looks swollen, but then a loop of bowel appears, rapidly followed by more until most of the intestines are outside the abdomen.

- If the intestines are not dirty or damaged, wrap the lamb in a clean towel, phone the surgery to warn of your emergency and take straight to your vet. It may be possible to replace the intestines, although the lamb will have to be anaesthetised in order to do this.

- If the intestines are very cold or damaged, it is kinder to humanely kill the lamb as soon as possible.

Atresia ani

This is an occasional developmental abnormality, where the anus is absent and the lamb is therefore unable to pass any faeces. Although an affected lamb may be noticed fairly soon, if, for example, someone attempts to take its temperature, most are not observed for several days until the belly becomes swollen and the reason then becomes obvious. This is another job for your vet, who may be able to do a small operation to cut into the rectum and stitch the lining to the skin to form an opening. The operation is actually easier to perform when the lamb is a few days old rather than newborn, since the build-up of faeces helps in finding the blind end of rectum. Occasionally, the gut ends further forward in the abdomen, though this will not be apparent until the operation is attempted. In this case nothing can be done so the lamb will have to be put down.

Entropion

In some flocks entropion (inturned eyelids), which is present at birth and affects the lower eyelid, is quite common. This is probably an inherited defect, although the mode of inheritance is not straightforward. Nevertheless, if it can be linked with the introduction of a new ram, he should be culled to try to eliminate the problem. If it does occur, you should check all lambs at birth, as it can usually be easily corrected at this stage. If it is missed, the eye becomes damaged and correction is much more difficult.

- Gently pull down the lower lid of each eye in turn and look for the margin of the eyelid (Figure 19.2a). In affected lambs, the lower lid of one or both eyes rolls inwards, so that the hairs rub on the surface of the cornea (Figure 19.2b).

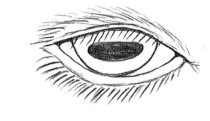

Figure 19.2a
Normal lamb's eye

Figure 19.2b
Eye affected with entropion
(inturned eyelid)

- Correct by simply pulling the eyelid out to the correct position and applying downward pressure on the skin below the eye (Figure 19.3).

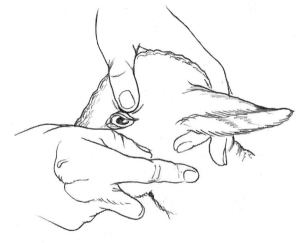

Figure 19.3
Correcting entropion by everting lower lid

- Recheck the lamb after an hour or two to make sure it has not recurred; if so, repeat the correction and keep rechecking, until you are happy it is staying in position.

If lambs affected with entropion are not treated soon after birth, the first thing which is usually noticed is that the eye(s) is runny and half-closed, and the cornea is white rather than clear. Although runny eyes can also result from eye infections, any lamb with this appearance should always be carefully examined to rule out entropion: pull down the lower eyelid to make sure that you can see the rim of the lid. If it is rolled in, correction is more difficult at this stage. There are various methods using injections into the eyelid, metal clips or stitching, but you should not attempt them unless you have been instructed by your vet. Inexpert use of needles near the eye can result in serious damage and blindness if you don't know what you are doing. Not correcting entropion can also result in blindness, so this condition should never be neglected.

Contracted tendons

Lambs are occasionally born with bowed front legs or contracted tendons, which cause them to stand on the very tip of the toes, or with the fetlock joints bent over so that they walk on the front of the fetlocks.

- If the lamb is managing to stand on the tips of the toes, the problem will usually correct itself within a day or two when the weight of the lamb causes the legs to return to the correct position.

- If the lamb is unable to straighten the legs, they should be dealt with immediately whilst the legs can be manipulated more easily than later. An easy and usually effective method is to apply splints made of plastic piping of an appropriate diameter. Carefully pad the leg with cotton wool before the splint is put on with the bottom end level with the bottom of the toe and the top end above the knee. Check to make sure that the top and bottom edges of the splint

are padded and will not cut into the leg, then fasten on firmly but not too tightly with an adhesive bandage. The splint should be checked every day and removed as soon as the lamb is able to stand on the toe (usually 3 to 4 days), as the leg can be damaged if the bandage becomes too tight.

Infections causing scouring, joint ill, navel ill and liver abscesses

These infections are most common in intensive lambing conditions, both indoors and outdoors, where large numbers of lambs are being born in a confined area. Scrupulous attention to detail, including early colostrum intake, hygiene of lambing pens and navel treatment, are important management factors in preventing these types of disease from occurring. Health of ewes is also important, specifically including foot health, since sheep affected with footrot are responsible for contamination of pens with bacteria which can particularly infect the navels of lambs.

If outbreaks of disease do occur, you will need veterinary help to identify the particular organisms involved and to decide appropriate treatment and specific preventive measures.

An emergency rehydration mixture can be made as follows.

Use a 5 ml plastic medicine spoon for measuring. Take 1 **level** 5 ml spoonful of salt (this is about 10 g). Remove a little so that about 9 g remain. Add this to 1 litre of water for a stock saline solution. When needed for feeding a scouring lamb, take 50 ml, warm to blood heat and add 1 heaped 5 ml spoonful of glucose powder (this is about 5 g). Feed with a bottle or stomach tube, re-peating every hour until the lamb looks brighter. If it has not improved within a few hours, consult your vet.

Diseases affecting the brain, nerves and muscles

A number of diseases (such as swayback, border disease, daft lamb disease, nutritional myopathy) affect the normal development of brain, spinal cord or muscles of the foetus depending on the severity, in such a way that lambs die straight after birth, are unable to stand properly, or may be mentally abnormal. The cause may be genetic, infectious or nutritional. Making a correct diagnosis is the starting point: consult your vet if this sort of pattern is emerging.

Dead lambs

Inevitably there will be some lamb deaths. It is useful to keep records and to try to decide at what stage the lamb died (before birth began, during the birth process or after birth) so that if losses are abnormally high the likely cause can be determined. Your vet will give you more information, but here is a basic guide.

Before birth

- Lambs which died a long time before birth are usually mummified (dried and shrivelled-up, often dark brown in colour).

- Lambs which died a day or two before birth show an opaqueness or whiteness of the corneas of the eyes.

- The lamb may appear smaller than usual and may have a poorly developed fleece.

These signs may all be associated with the presence of one of the infectious causes of abortion, so steps should be taken to check for these (see page 20). Lambs may also die in the womb if the ewe is carrying a large litter and the placentas are not able to sustain the development of all the lambs. They may also die if the ewe is ill with, for example, twin lamb disease (pregnancy toxaemia).

During birth

- These are lambs which are born at the correct time and are fresh looking.

- The eye is usually clear, but the lambs are dead, or die immediately after birth.

- Part of the lamb such as the head, tongue or a leg may be swollen.

Here something has gone wrong during the birth process – perhaps the lamb was too large, or presented wrongly, or for some reason the ewe did not get on with the job properly. If a part of the lamb is swollen, this shows that there was a malpresentation or that the birth was too prolonged.

After birth

- Lambs which are smaller than normal, and very large lambs, are more likely to die than those of a normal size.

- Small lambs die because they do not get enough colostrum or lose heat rapidly from their relatively large body surface area, so they get hypothermia (see page 77) and are prone to infectious diseases.

- Large lambs are often slow to be born and suffer from a lack of oxygen which may affect the brain; this causes 'doziness', lack of 'get up and go' and problems with body temperature control (see 'Resuscitation', page 64).

CHAPTER 20

Conclusion

We hope that by this stage you have healthy ewes and lambs ready for the next phase of growth and development. It is difficult to know where to draw the line, but we have decided not to proceed any further in this book. If you have sick ewes or lambs you will need to obtain help and advice from your vet.

As we said at the beginning, lambing time cannot be taken in isolation. A successful outcome depends on year-round planning, ensuring that the ewes are in the optimum condition at each crucial part of the production cycle, an appropriate vaccination schedule, maintenance of a high standard of hygiene in lambing sheds and pens, conscientious observation and clean, gentle hand-ling when ewes and lambs require help. We hope that this book will contribute to a reduction in the losses of both ewes and lambs which occur every year around lambing. For further information, particularly about other parts of the sheep breeding cycle, we recommend the following books:

Clarkson, M.J. and Winter, A.C. (1997), *A Handbook for the Sheep Clinician,* 5th Ed., Liverpool University Press, UK.

Henderson, D.C. (1990), *The Veterinary Book for Sheep Farmers,* Farming Press, Ipswich, UK.

Henderson, D.C. (1993), *Lamb Survival* (video), Farming Press, Ipswich, UK.

Winter, A.C. and Charnley J.G. (1999) *The Sheep Keeper's Veterinary Handbook*, The Crowood Press Ltd, Ramsbury, Marlborough, Wiltshire, UK.

Index